WYOMING STATE GEOLOGICAL SURVEY
Gary B. Glass, State Geologist

Report of Investigations No. 53

DIAMONDS AND MANTLE SOURCE ROCKS IN THE WYOMING CRATON WITH A DISCUSSION OF OTHER U.S. OCCURRENCES

by

W. Dan Hausel

Laramie, Wyoming

1998

Table of contents

Illustrations

Figures

Tables

Plates

Abstract

Many diamonds found in the United States were derived from known host rocks (e.g., kimberlite and lamproite). The distribution of the known diamondiferous host rocks as well as the associated basement terrane indicates that the greatest potential for the discovery of commercial diamond deposits in the U.S. lies within the Wyoming craton. To date, more than 100 kimberlites, one of the largest lamproite fields in the world, and more than 130,000 diamonds, have been found along the margin of this craton.

Several secondary diamond occurrences have also been found in the United States. Many of these may have been derived from undiscovered kimberlites and lamproites; however, the hundreds of diamonds reported in the Pacific Coast region lie in an area considered to be unfavorable for kimberlite and lamproite. It is possible that many of these diamonds originated from unconventional host rocks.

Introduction

Diamonds have been found in the United States ranging from small crystals less than 1 carat to thumbnail-size crystals weighing up to 40 carats. The source of many of these diamonds has remained a mystery; however, with the relatively recent identification of several diamondiferous host rock sources, one can deduce that many of these diamonds were derived from the erosion of yet-undiscovered ultramafic to mafic host rocks. Potential host rocks have now been identified in many states within the U.S.

Diamonds have been recovered from several kimberlites in the Colorado-Wyoming State Line district; possibly from a cryptovolcanic breccia pipe in the Green River Basin, Wyoming; from kimberlites in Michigan; and from lamproites in Arkansas. Diamonds were reported (although not verified) in kimberlite in Kansas, and in peridotite in New York and Maryland. Secondary (not in-place) diamond occurrences are widespread in the United States, and diamonds have been recovered from gravels in numerous gold placers and from glacial till. Evidence now suggests that North America may someday become a major source of diamonds, and could possibly become the world's largest producer of the gemstone.

In this report, the primary and secondary diamond occurrences in the United States are separated into the following six geographic regions: Appalachian Mountains, Great Lakes, Continental Interior, Gulf Coast, Great Basin-Rocky Mountain, and Pacific Coast. Most notable of the secondary occurrences are the Appalachian Mountains and Pacific Coast regions where hundreds of diamonds, including some of the largest diamonds reported in the United States, have been found (**Figure 1, Table 1**). The source of most of these placer diamonds is unknown in that a large part of these terranes is not considered favorable for diamondiferous kimberlite or lamproite.

The principal primary diamond deposits in the United States are located in the Colorado-Wyoming State Line district and the Murfreesburo, Arkansas area. More than 130,000 diamonds have been produced from several kimberlites in the State Line district, and more than 90,000 diamonds have been produced from a group of lamproites in the Murfreesburo area. These two areas have yielded several of the largest diamonds found in the United States (**Table 1**).

Diamonds and the potential diamondiferous rocks kimberlite, lamproite, lamprophyre, and peridotite have been reported at a number of localities in the United States (**Figure 1**). These localities are discussed in the following sections of this report: (1) Wyoming craton, (2) Great Basin-Rocky Mountains exclusive of the Wyoming craton, (3) Appalachian Mountains, (4) Great Lakes, (5) Continental Interior, (6) Gulf Coast, and (7) Pacific Coast. A few of the potential "host" rocks are known to contain diamond, but many of the intrusives have not been sampled thoroughly.

Diamonds and their host rocks

Diamond is the hardest naturally occurring mineral on the surface of Earth. Diamonds have been found in a number of rock types throughout the world, but commercial diamond deposits are, for the

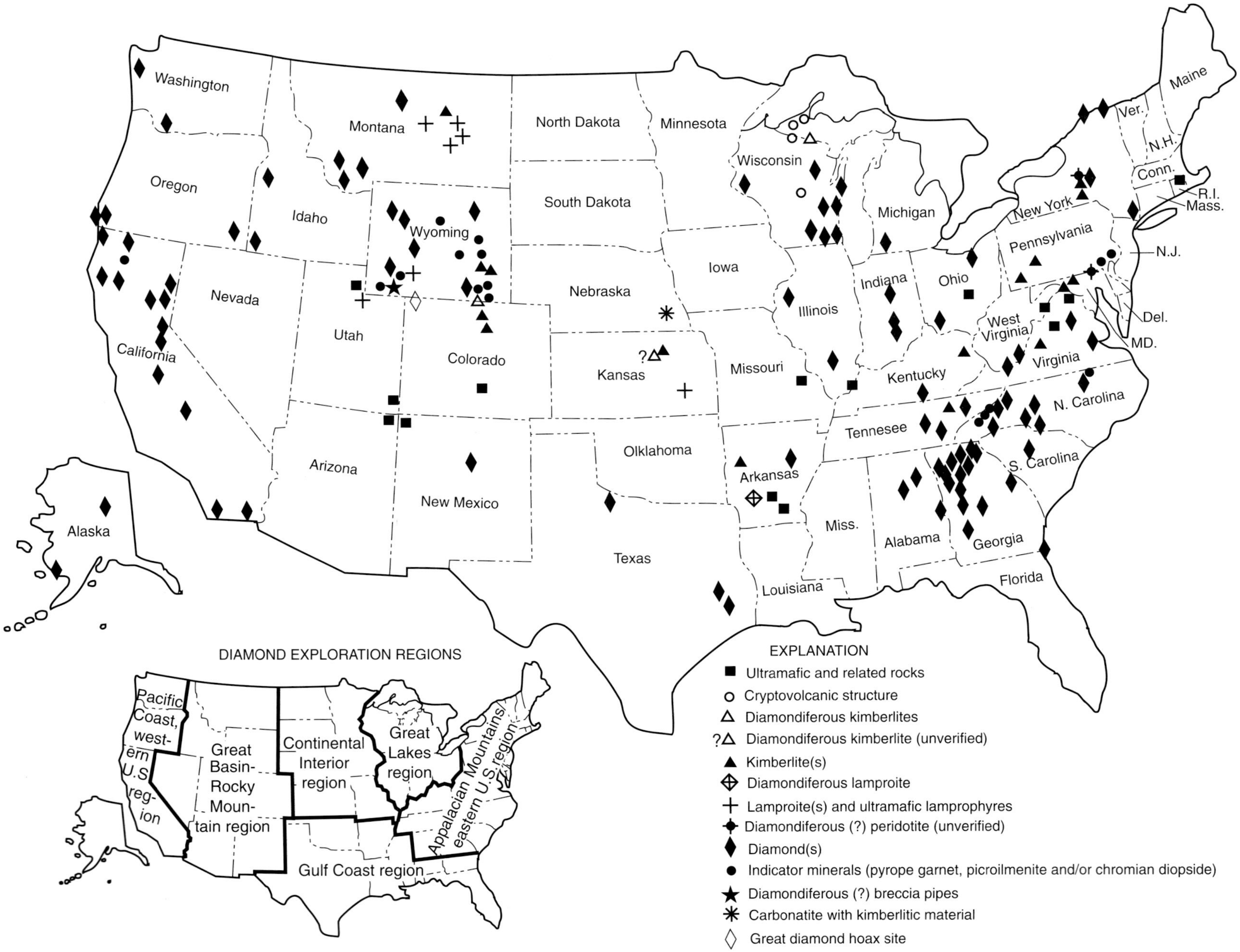

Figure 1. Diamond, kimberlite, lamproite, and related rock localities in the United States (each locality may represent one or more reported occurrence).

Table 1. Reported diamonds in the United States weighing more than 2 carats (modified from Hausel, 1995a, b).

Diamond	Weight (carats)	State	Diamond	Weight (carats)	State
Uncle Sam	40.42	Arkansas	Williams Ferry	>6.00	Georgia
Punch Jones	34.46	West Virginia	Milford	6.00	Ohio
Star of Murfreesburo	34.25	Arkansas	Unnamed	5.90	Arkansas
Doubledipity	32.99	California	Unnamed	5.76	Arkansas
Kelsey Lake	28.30	Colorado	Unnamed	5.63	Arkansas
Kelsey Lake	28.18	Colorado	Unnamed	5.58	Arkansas
Howell	27.31	Arkansas	Sloan	5.51	Colorado
Dewey	23.75	Virginia	Unnamed	5.19	Arkansas
Terresa	21.25	Wisconsin	Unnamed	5.15	Arkansas
Rock Flat	19.50	Idaho	Unnamed	5.08	Arkansas
Enigma	17.83	California	Unnamed	5.00	Arkansas
Amarillo Starlight	16.37	Arkansas	Unnamed	5.00	Arkansas
Kelsey Lake	16.29	Colorado	Stanley	4.88	Indiana
Eagle	15.37	Wisconsin	Lee	4.61	Alabama
Star of Arkansas	15.24	Arkansas	Shelby	4.37	Alabama
Serendipity	14.33	California	Dysortville	4.33	North Carolina
Kelsey Lake	14.20	Colorado	Daniel Light	4.25	Georgia
Lewis and Clark	14.00	Montana	Skamania	>4.00	Washington
Kelsey Lake	11.85	Colorado	Brown County	4.00	Indiana
Dowagiac	10.875	Michigan	Peru	3.93	Indiana
Kelsey Lake	10.48	Colorado	Devine	3.87	Wisconsin
Kelsey Lake	9.40	Colorado	Jeopardy	3.90	California
Unnamed	8.82	Arkansas	Columbus	3.50	Georgia
Unnamed	8.61	Arkansas	Chicken Craw	3.15	Oregon
Unnamed	7.95	Arkansas	Salt Creek	3.06	Indiana
Ashley	7.75	Illinois	Little Indian Creek	3.00	Indiana
French Corral	7.25	California	Union Crossroads	3.00	Tennessee
Shell Bluff	7.11	Georgia	Chicken Park	2.60	Colorado
Unnamed	6.75	Arkansas	Malheur	2.50	Oregon
Saukville	6.57	Wisconsin	Prescott Siding	2.41	Alabama
Gary Moore	6.43	Arkansas	McDowell	2.38	North Carolina
Unnamed	6.30	Arkansas	Gold Creek	2.28	Indiana
Unnamed	6.25	Arkansas	Mary	2.27	California
Unnamed	6.20	Arkansas	Moore	2.25	California
Kelsey Lake	6.20	Wyoming	George Creek	2.14	Colorado
Eisenhower	6.11	Arkansas	Burlington	2.11	Wisconsin
Unnamed	6.07	Arkansas			

most part, restricted to kimberlite, lamproite, and the secondary deposits derived from these primary host rocks. Diamonds have also been recovered from a number of unconventional host rocks (Helmstaedt, 1993; Janse, 1994a; Hausel, 1994b, 1996c). To date, unconventional deposits have been relatively unproductive.

Diamonds

There are two basic crystal forms of natural diamond: (1) isometric, and (2) hexagonal. Terrestrial diamonds, for the most part, are isometric. Some isometric microdiamonds are also derived from meteorites. Hexagonal diamonds (lonsdaleite) are rare, and appear to be restricted to meteorites. However, a rare deposit in Siberia, known as the Popigay Depression, was once mined for hexagonal diamonds. This anomaly is interpreted by most researchers as an impact structure, although a few researchers have suggested that it may represent a terrestrial diatreme, which sampled rocks deep within Earth's mantle (Ed Erlich, personal communication, 1986).

Isometric diamonds occur in a number of forms and habits. Isometric diamond, in its simplest form, produces six-sided cubes referred to as hexahedrons. The most common habit of diamond is the octahedron, or some modification of the octahedron (**Figure 2**). Octahedrons form eight-sided bipyramids, although some octahedrons may develop ridges on the octahedral faces producing crystals of trisoctahedral or hexoctahedral habit. Partial resorption of octahedral diamonds can also result in a rounded dodecahedron (12-sided) with rhombic faces. Many dodecahedrons develop ridges

Figure 2. Octahedral diamond with distinct 111 (triangular) crystal face. This 14.2-carat gemstone was recovered from the Kelsey Lake mine along the Colorado-Wyoming border (photograph courtesy of Redaurum Ltd. and H. Coopersmith).

on the rhombic faces producing a 24-sided crystal known as a trishexahedron. Additionally, four-sided tetrahedral diamonds are sometimes encountered; these are probably distorted octahedrons (Bruton, 1979).

Diamonds often enclose mineral inclusions, which provide important scientific data on the origin of diamond; e.g., inclusions are used to date the formation of the diamond. The diamonds that have been analyzed to date have yielded Precambrian ages [3.3 Ga (Giga-annum or billions of years old) to 630 Ma (Mega-annum or millions of years old)], with the majority of them occurring within the period of about 3.2 to 2.4 Ga (Anonymous, 1994). One exception, however, is a group of diamonds from an apparently unconventional source rock in New South Wales, Australia. These diamonds yielded age dates of only 300 Ma (S.R. Lishmund, personal communication, 1994).

It has been determined that the conventional host rocks for diamonds are significantly younger than the diamonds themselves. For example, most lamproite and kimberlite "host" rocks yield much younger ages (1.6 Ga to 1.0 Ma) than the diamonds themselves (Helmstaedt, 1993). This age discrepancy supports the hypothesis that most diamonds are xenocrysts rather than phenocrysts, and were accidentally trapped in kimberlite or lamproite magma deep within Earth's upper mantle.

Mineral inclusions in diamond are characteristic of peridotite or eclogite source rocks. These diamonds are designated as P-type (peridotitic) or E-type (eclogitic). Peridotites are ultramafic rocks that contain pyroxene and more than 40% olivine, and are subdivided based on the type of pyroxenes present (**Plate 1a**). For example, lherzolites are peridotites that contain clinopyroxene and orthopyroxene; wehrlites contain clinopyroxene but lack orthopyroxene; and harzburgites contain orthopyroxene but lack clinopyroxene. Thus, P-type diamonds may contain inclusions of olivine, orthopyroxene, clinopyroxene, garnet, chromite, diamond, and/or sulfides, all derived from their peridotitic host rock.

Some peridotitic garnets have unique chemistries indicative of the diamond stability field. Since garnets are typically more abundant than diamonds (especially in kimberlite), they are often used in the exploration for diamondiferous deposits. Some peridotitic pyrope garnets have been designated as "G10," which have been derived from sub-calcic (Ca) harzburgites, and "G9," derived from the more calcic lherzolites (Gurney, 1986). Most garnet inclusions in diamonds are derived from sub-Ca harzburgites. These garnets have relatively low Ca/Cr ratios compared to lherzolitic pyropes (pyrope garnets) (Gurney, 1989). Thus, garnets found within a host kimberlite or lamproite that have similar chemistry to diamond-inclusion garnets, are interpreted to have originated within the diamond stability field.

Eclogite, another common mantle xenolith found in kimberlite, has an approximate chemical composition of basalt. Mineralogically, eclogites consist of omphacitic clinopyroxene and almandine-pyrope garnet, and may have accessory rutile, kyanite, corundum, coesite, and possibly diamond. Thus the mineral inclusions found in E-type diamonds are characteristic of eclogite.

Some eclogitic garnets also have unique chemistries indicative of the diamond stability field. Mantle garnets of eclogitic parentage have been designated as either Group I or Group II. The Group I eclogitic pyrope garnets have Na_2O contents of ≥0.09 wt.% and low levels of Cr_2O_3 (typically <0.05 wt.% Cr_2O_3). All diamond-bearing eclogites belong to Group I. Group II eclogitic pyrope garnets are similar to the Group I garnets, but have <0.09 wt.% Na_2O. Eclogitic garnets have CaO contents in the range of 3.5 to 20 wt.%. Similar, low-Cr garnets that contain less than 3.5 wt.% CaO are probably derived from crustal rocks (Schulze, 1993).

Along with garnet, some other indicator minerals are used to provide information on the diamond potential of kimberlite or lamproite. These include chromite, clinopyroxene, picroilmenite, etc. (see McCallum and Waldman, 1991; McCandless and Gurney, 1989; Fipke and others, 1995).

Some nodules (or mantle xenoliths) recovered from kimberlite have been very rich in diamond. For example, a hand-specimen of eclogite collected from the Sloan 1 diatreme in the Colorado-Wyoming State Line district yielded a minimum diamond grade of 2,100,000 carats/100 tonnes (metric tons) (Schulze, 1992). Commercial kimberlite and lamproite diamond ore may only average 5 to 680 carats/100 tonnes.

Kimberlite

Kimberlite is considered one of the principal ores for commercial quantities of diamond. Since the great majority of diamonds found in kimberlite (as well as lamproite) are considered xenocrysts, kimberlite cannot be considered as true host rock, as the kimberlite acts only as a transportation medium for the gems. As a result, not all kimberlites contain diamond. Typically, only about 10% are diamondiferous and less than 2% have commercial amounts of diamonds in concentrations less than 1 ppm (parts per million) (Lampietti and Sutherland, 1978).

Kimberlite is a volatile-rich, potassic, ultrabasic hybrid igneous rock with variable mineralogy that forms pipes, dikes, and blows (**Figure 3**) (McCallum and Mabarak, 1976; Kjarsgaard, 1996). Blows swell from dikes and are considered the root zones of the pipes. In cross section, the classical model of a kimberlite pipe is a carrot-shaped structure that originates from a dike upward into a root that expands into a diatreme and may erupt at the surface forming a crater. The root zone may develop as deep as 1.5 miles below the surface, forming an irregular dike complex occupied by one or more intrusive phases of hypabyssal facies kimberlite.

The hypabyssal facies kimberlite is massive, porphyritic kimberlite with little to no breccia fragments or country rock clasts. Diatreme facies kimberlite is represented by volcaniclastic breccias consisting of clasts of country rock, fragments of hypabyssal kimberlite, and pelletal lapilli in a matrix of serpentine and diopside (**Plates 1b** and **1c**). Crater facies kimberlite, consisting of pyroclastic and epiclastic kimberlitic material (pyroclastics, tuffs, and lapilli pyroclastics), is preserved only in pipes that have undergone little or no erosion since emplacement (crater facies kimberlite is rare, as most pipes have been subjected to erosion). Where mineralized, the grades of crater facies kimberlite are often significantly higher than the diatreme facies (Helmstaedt, 1993), and ore grades may decrease with depth in the pipe (Gurney and others, 1991). Although the grades of diatreme facies and root zone kimberlite may not show any noticeable differences, different intrusive phases of kimberlite in the same pipe often yield different diamond grades (Wagner, 1914).

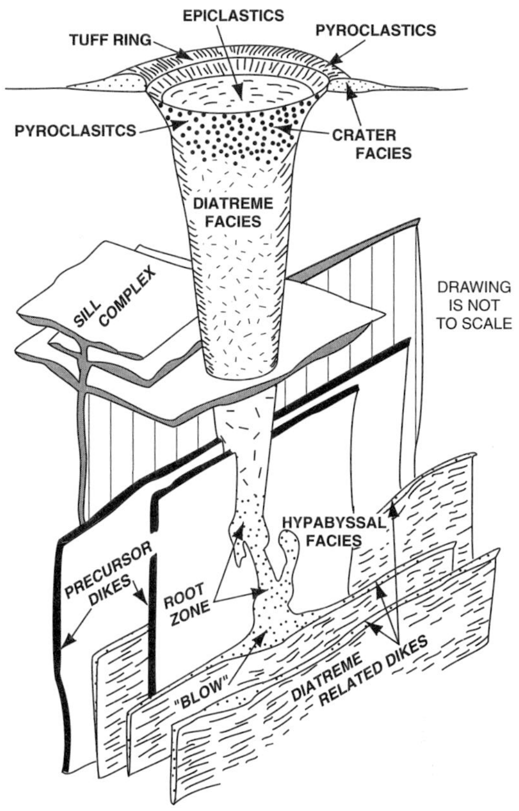

Figure 3. Model of a kimberlite pipe showing crater facies, diatreme facies, and hypabyssal facies kimberlite (not to scale) originating from a root zone at depth (from Mitchell, 1986).

Two varieties of kimberlite were recognized by Wagner (1914): basaltic (Group I) kimberlite and micaceous (Group II) kimberlite. Because of their unique chemistry and mineralogy, Mitchell (1995) suggested that the Group II kimberlites–which have closer affinities to lamproite–be renamed orangeites based on their original discovery in the Orange Free State, South Africa.

Lamproite

Diamondiferous lamproite was recognized in 1978 in the Kimberley Province of Western Australia (**Figure 4**). Before this discovery, kimberlite was assumed to be the only commercial primary source rock for diamond, even though diamonds had been mined from lamproites in Arkansas in the early 1900s and in India in the early 1800s. However, these lamproites were thought to be kimberlites until very recently. Kimberlites and lamproites are distinctly different rock types, although there is an

Figure 4. The Argyle mine and mill in Western Australia as seen in 1986. The Argyle olivine lamproite is extremely diamond rich, yielding ore grades as high as 680 carats/100 tonnes.

overlap in the chemical compositions of micaceous kimberlite and lamproite (Helmstaedt, 1993; Peterson, 1996).

In general, lamproites are peralkaline ultrapotassic mafic igneous rocks enriched in the trace elements Zr, Nb, Sr, Ba, and Rb relative to kimberlite. Lamproite, like kimberlite, is silica-poor and rich in MgO, FeO, K_2O, and volatiles (Kirkley and others, 1991). Lamproites may contain diopside, phlogopite, K-Ti richterite, leucite, sanidine, wadeite, priderite, and/or olivine, with minor apatite, perovskite, ilmenite, and spinel (Mitchell and Bergman, 1991). The typical kimberlitic indicator minerals pyrope garnet, chromian diopside, and picroilmenite are uncommon in lamproite.

Kimberlitic indicator minerals are minerals with a unique chemistry that provide a signature from Earth's upper mantle. These minerals are typically associated with kimberlite (hence the name "kimberlitic indicator minerals" as used in this report), but they can also occur in any ultramafic to mafic igneous rock which has had a similar origin in the upper mantle. For instance, these kimberlitic indicator minerals have also been found in some mantle-derived lamproites and ultramafic lamprophyres.

Lamproite occurs as extrusive, subvolcanic, and hypabyssal facies rock, and erupts from small volcanoes with restricted flows. Where mineralized, diamond xenocrysts in lamproite are primarily restricted to the pyroclastics; the magmatic phases (and sills) are notoriously diamond poor. Thus, the available minable tonnage is primarily limited to the vent facies rocks. Typically, diamond grades are higher in olivine-lamproites than in the leucite-lamproites.

The known diamondiferous lamproites in the world are found at Argyle and in the Ellendale field of Western Australia; Kapamba in Zambia; Majhgwan-Chelima in India; Murfreesburo, Arkansas; Aldan in Russia; and Bobi in the Ivory Coast. In that some petrologists now recognize Group II kimberlites as a variety of lamproite (orangeite), lamproites appear to be a very important source of diamonds: many Group II kimberlites have been a significant source of gem diamonds (Peterson, 1996). In addition, the Argyle olivine lamproite in Australia yields more diamonds per tonne (680 carats/100 tonnes) than any other primary diamond deposit in the world. Currently, the Argyle mine produces about 30% of the world's diamonds, with some rare pink diamonds that recently became the most expensive gemstones in the world. Some "Argyle pinks" have been valued at about $1 million (Australian) for a 1-carat brilliant cut (Rock and others, 1992) (**Plate 1d**).

Altered olivine-phlogopite lamproite dikes have also yielded diamonds from the Ivory Coast and Gabon, in Western Africa. These rocks are devoid of the usual kimberlitic indicator minerals and appear as talc or phlogopite schists. The Bobi lamproite dike in the Ivory Coast is locally very diamond rich with grades up to 1000 carats/100 tonnes (Helmstaedt, 1993).

Ultramafic complexes, lamprophyres, and related rocks

Along with kimberlites and lamproites, diamonds have been found in a variety of other "host" rocks that are considered unconventional; as such, these diamondiferous rocks are currently considered to be scientific curiosities. For example, diamonds have been reported in lamprophyres, alkali basalts, and other unconventional host rocks.

Certain types of ultramafic lamprophyres, notably alnöites, have yielded minor amounts of diamond. Alnöites lack the characteristic kimberlitic indicator minerals, contain melilite and biotite mica (instead of phlogopite), and can be thought of as melilite-bearing, mica-peridotites (Helmstaedt, 1993).

Diamondiferous picritic monchiquites have been reported in Western Australia, and diamonds have also been reported in lamprophyres in Quebec, Canada. Diamonds were also reported in mugerarite, nephelinite, and alkali basalt from New South Wales, Australia (Jaques and others, 1989; Engineering

Plate 1

Plate 1a. Garnet peridotite nodule collected from the Schaffer group of kimberlites in the Colorado-Wyoming State Line district. The peridotite contains rounded, reddish, pyrope garnet porphyroblasts with green chromian diopside (a clinopyroxene) in a serpentinized matrix. Much of the serpentine forms replacements after olivine.

Plate 1b. Hand specimen of diatreme facies kimberlite from the Sloan 1 intrusive, Colorado-Wyoming State Line district. Clasts of country rock including a reddish pyrope megacryst are seen in the breccia.

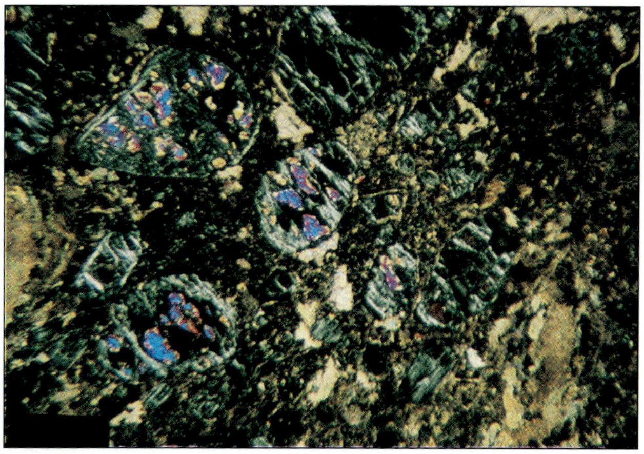

Plate 1c. Thin section of Nix 1 kimberlite under polarized light shows serpentine pseudomorphs after olivine with some remnant olivine.

Plate 1d. A collection of colored gemstone diamonds including Argyle pinks from the Argyle mine, Western Australia.

and Mining Journal, 1994). A set of five diamonds from this region in New South Wales yielded a relatively young age date of 300 Ma (S.R. Lishmund, personal communication, 1994). These diamonds are assumed to have formed at relatively shallow depths (48 miles) in a cool, subducted, organic-rich, oceanic slab (Engineering and Mining Journal, 1994).

Other primary diamond deposits that have recently been recognized are high-pressure alpine-type and ophiolitic peridotites. Such peridotites were discovered in the Ural, Caucasus, and Koryak Mountains of the former Soviet Union after diamonds were traced from nearby placers. The initial discovery occurred in 1978, and resulted in identifying diamonds in harzburgite bedrock in the Phanerozoic Koryak Mountains of northern Kamchatka (Janse, 1994a). Diamonds in alpine peridotites from the Koryak Mountains, as well as from Tibet, are thought to have formed during subduction and to have survived metastably during rapid tectonic uplift.

Diamonds have also been found in a metamorphosed, layered, mafic-ultramafic complex near Kaya, Burkina Faso, Western Africa. This particular complex consists of metamorphosed dunite overlying a sequence of layered amphibolites, metaperidotites, metapyroxenites, and biotite-plagioclase gneiss (Helmstaedt, 1993).

Recently, a slab of oceanic lithosphere from a high-pressure metamorphic belt near the Strait of Gibraltar in Morocco was shown to have been diamondiferous prior to emplacement. The Beni Bousera peridotite massif contains numerous graphite pseudomorphs after diamond in garnet clinopyroxenite. The mineral inclusions in the graphite, and the carbon isotope composition of the graphite, are similar to E-type diamonds from kimberlite.

The graphite octahedra in the Beni Bousera massif are confined to four garnet clinopyroxenite magmatic cumulate layers in the complex. Two of the layers are greater than 6 feet thick, contain orange pyrope-almandine garnet with compositions comparable to those found in diamond-bearing eclogites (Na_2O concentrations up to 0.14 wt.%), and contain omphacitic pyroxene porphyroclasts with minor plagioclase, spinel, and sulfides. The graphite-bearing garnet clinopyroxenite layers, along with wehrlites, lherzolites, and diopsidites, form an intercalated horizon up to 50 feet thick at the apex of the massif (Pearson and others, 1989, 1993).

Although primary diamond was not preserved in the Beni Bousera slab, the concentration of the graphitized diamonds indicates the slab initially contained about 15% diamond, or approximately 10,000 times as many diamonds per unit mass of rock than any known kimberlite intrusive (Pearson and others, 1993). This deposit may not be unique in that many detrital diamonds have been reported along plate margins near ophiolite complexes elsewhere in the world including the west coast of the United States (Hausel, 1996c).

High pressure metamorphic rocks

Diamonds have been found in some metamorphic environments. For instance, microdiamonds (averaging only 12.5 microns in length) were found as inclusions in garnets and zircons in metasedimentary garnet-biotite gneisses and schists in the Kokchetav massif of northern Kazakhstan. In addition to these unusual occurrences, diamonds have also been identified as inclusions in garnets in coesite-bearing eclogites, garnet pyroxenites, and jadeites in the Dabie Mountains of eastern China. These latter diamonds average only 10 to 60 microns across and include grains up to 240 microns (Helmstaedt, 1993).

Metamorphic diamonds were also discovered in 1992 northwest of the Ulaan-Baatar region, Mongolia, by the Central Institute of Exploration from Moscow. Initial bulk sampling tests recovered diamonds in the range of <0.1 mm (millimeter) to 1 mm in diameter from the metamorphic rock, which yielded grades from 4000 carats/100 tonnes to 10,000 carats/100 tonnes. This deposit is similar to the Kokchetav deposit in Kazakhstan (Ed Erlich, written communication, 1993).

Meteorites

Diamonds occur in some meteorites. As much as 1% diamond (both isometric and hexagonal) by weight has been found in stoney and iron meteorites. It is suggested that many microdiamonds found in the sedimentary record are derived from meteorites. Lonsdaleite (the hexagonal diamond form) has also been reported from the Popigay Depression in northern Siberia. In the past, diamonds were mined from the Popigay Depression for industrial purposes because of their extreme hardness. The Popigay structure has been described as an astrobleme, although some researchers have suggested this structure may have had terrestrial origins (Ed Erlich, personal communication, 1985).

Placers

Isometric diamonds are 6000 to 8000 times harder than any other mineral. Because of their extreme hardness and chemical inertness, diamonds can be transported in streams over great distances. In some instances, diamonds are assumed to have been transported hundreds of miles from their original source rock (e.g., Gurney and others, 1991).

Favorable terranes for diamonds

According to current diamond exploration models, diamonds transported to the surface by kimberlites or lamproites originated from peridotite or eclogite at depths of possibly 90 to 120 miles within the upper mantle. In order to produce magmas at these depths, it is necessary to have thick, cool, lithospheric root zones. Such lithospheric roots are interpreted to exist under Archean cratons, which are ancient continental cores older than 2.5 Ga (i.e., Kirkely and others, 1991; Levinson and others, 1992). Such cratons have attained stability, and have been subjected to very little deformation over a very long period of time, typically more than 1.5 billion years.

However, within the past 20 years, diamondiferous kimberlite and lamproite have been recognized in Proterozoic mobile belts along the margins of some Archean cratonic cores (i.e., Argyle, Australia; State Line district, Colorado-Wyoming). These diamondiferous intrusives were emplaced in Proterozoic basement rocks that were cratonized and attained stability between 2.5 and 1.6 Ga. Because of these discoveries, such cratonized Proterozoic terranes are now thought to have considerable economic potential.

Located along the edge of some of these Proterozoic cratonized terranes in North America are Proterozoic basement rocks which obtained stability at a much later time in geological history, sometime between 1.6 Ga to 800 Ma. Both kimberlite and lamproite have been found in these younger Proterozoic terranes; however, these intrusives are generally barren or only contain trace amounts of diamond (Kirkley and others, 1991). These Proterozoic belts are currently considered poor exploration targets for diamond.

Thus, the cratonic areas of the world can be subdivided into favorable and less favorable terranes for diamond exploration. These terranes were described by Janse (1994b) as (1) archons, which consist of Archean basement last subjected to a thermal event with a minimum age of 2.5 Ga; (2) protons, which consist of Early to Middle Proterozoic basement last subjected to a thermal event with a minimum age of 1.6 Ga; and (3) tectons, which consist of Late Proterozoic basement last subjected to a thermal event with a minimum age of 800 Ma. It is unfortunate that proton was used to describe one of these terranes, as it has obviously been used extensively in nuclear physics and chemistry. Possibly "Proterozoic A" and "Proterozoic B" would have been more useful terms than proton or tecton.

It should be noted that the largest cratonic region of the world lies under portions of North America (**Figure 5**). Even though North America has been only a very minor source of diamonds in the past, this is anticipated to change dramatically in the near future.

Besides the cratonic regions of the world, diamonds have also been found off-craton in association with unconventional source rocks in accreted oceanic crust adjacent to subduction zones along some plate margins. Research on these deposits is lacking as no commercial deposit has yet been found in these rocks. However, a considerable number of placer diamonds have been found in these terranes worldwide, leaving one to suspect that they may offer some economic potential either from diamondiferous obducted oceanic mantle (eclogite, peridotite, serpentinite) or from the intrusives that sampled these rocks at depth. For exploration purposes, these terranes are herein termed "plate margin terranes."

The Wyoming craton

The Wyoming craton consists of an Archean basement (archon) known as the Wyoming Province (**Figure 5**) and cratonized Proterozoic rocks (proton) of the Colorado Province that lie along the southern margin of the Wyoming Province. This craton underlies Wyoming and extends into the adjacent states of Montana, Colorado, northern Utah, extreme northeastern Nevada, and eastern Idaho (**Figure 6**).

Much of Wyoming and Montana are underlain by the Wyoming Province, which forms the core of the Wyoming craton. The Wyoming Province is bounded along the south by cratonized Proterozoic basement composed of volcanogenic gneiss and schist of the Colorado Province, and along the east by Archean-Proterozoic basement rocks of the Trans-Hudson orogen (**Figure 5**), which stabilized in the Proterozoic (Goldich and others, 1966; Hausel and others, 1991). The Trans-Hudson orogen is interpreted as a Proterozoic fold belt that welded the Wyoming Province on the west, to the Superior Province to the east (Williams and others, 1986; Eggler and others, 1988).

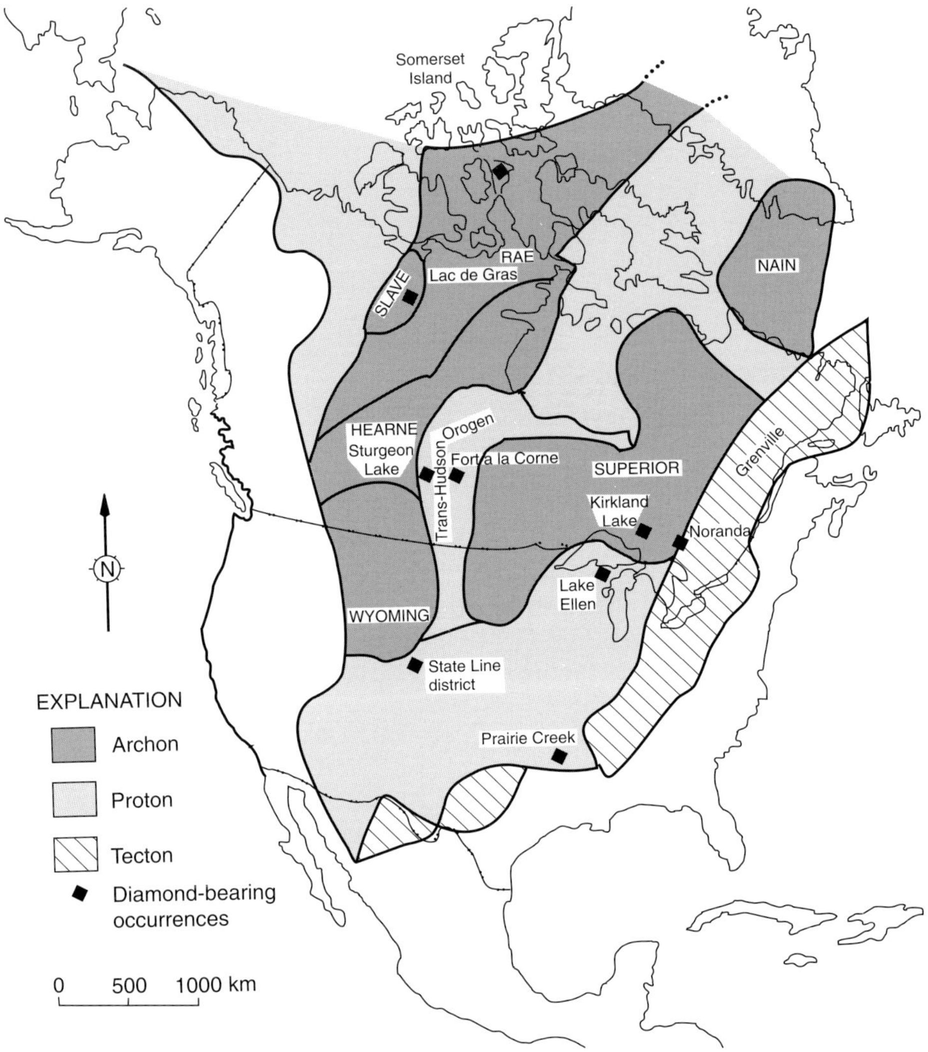

Figure 5. The North American craton showing regions of favorability for diamondiferous kimberlite and lamproite (from Levinson and others, 1992). Major Archean provinces are labeled in all capital letters.

The Wyoming Province is formed of Archean granite, gneiss, and supracrustal metamorphic rocks exposed in the core of several Laramide uplifts and present under the sedimentary basins between the uplifts, where it is hidden under a thick blanket of Phanerozoic sedimentary rocks.

According to Eggler and others (1988), the Wyoming craton was established by 2.7 Ga, but was later affected by a regional metamorphic event recorded at 1.9 to 1.7 Ga. The southern boundary of the Wyoming Province exposed in the Medicine Bow Mountains and Sierra Madre in southeastern Wyoming is known locally as the Mullen Creek-Nash Fork shear zone. This shear, which forms part of the Cheyenne belt suture, represents a continental-arc collision zone (Graff, 1978; Hills and Houston, 1979) separating the Wyoming Province to the north from cratonized (1.7 Ga) Proterozoic basement of the Colorado Province to the south. The Colorado Province south of the suture consists of volcanogenic island-arc basement rocks and intrusive granites.

The boundary of the Wyoming Province continues eastward from the Medicine Bow Mountains into the central Laramie Mountains, where it is masked by a 1.5-Ga anorthosite-syenite batholith. Sheared rocks in the Richeau Hills along the eastern flank of the anorthosite complex are thought to be part of the Cheyenne belt. From the Laramie Mountains, the boundary continues under Phanerozoic sedimentary rocks of the High Plains, and is projected to intersect the southward extension of the Trans-Hudson orogen near the Wyoming-Nebraska border.

The eastern boundary of the Wyoming Province is not well exposed, and in the past has included the Black Hills of South Dakota. Geochronologic data show that ages of the basement in western South Dakota have been largely reset to Proterozoic values of the Trans-Hudsonian event (Goldich and others, 1966). Isolated ages greater than 2.5 Ga in the Black Hills represent envelopes of an Archean basement that were thermally insulated during the Trans-Hudson orogeny, and their isotopic compositions were not reset. Thus, the eastern boundary of

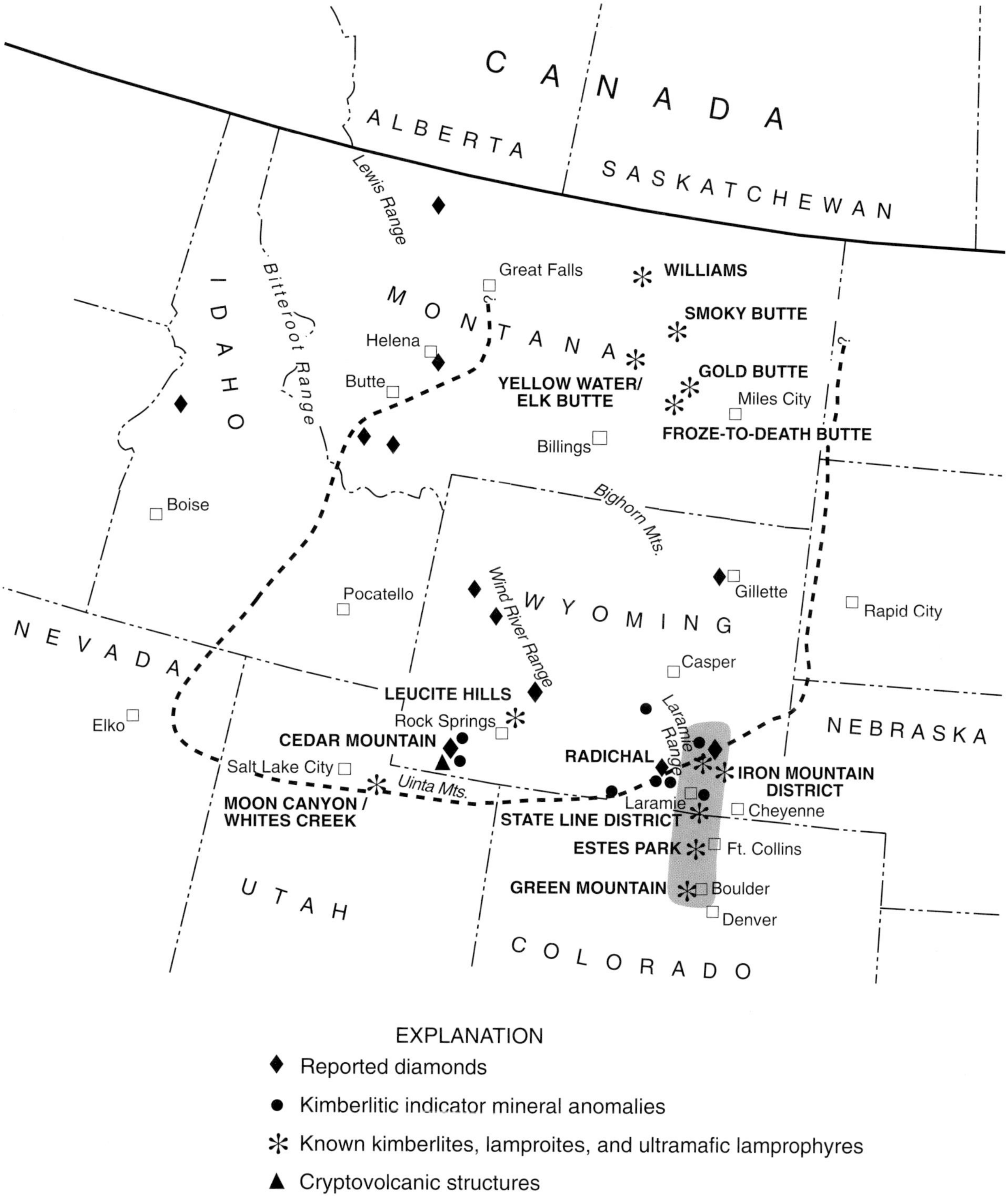

Figure 6. Generalized map of the Wyoming craton showing the Wyoming Archean Province (outlined by heavy dashed line) and locations of known kimberlites, lamproites, related rocks, and indicator mineral anomalies. The general location of the Colorado-Wyoming kimberlite province, which lies within the cratonized Proterozoic terrane of the Colorado Province, is shown by the shaded area.

the Wyoming Province essentially parallels the Wyoming-South Dakota border (Hausel and others, 1991; Sims, 1995).

From southeastern Wyoming, the southern boundary of the Wyoming Province continues westward through northern Utah and into the Basin and Range of extreme northeastern Nevada, where relict Archean ages were reported in the core of the Eastern Humbolt Range (Lush and others, 1988). From Nevada, the province is interpreted to continue northward under sedimentary cover in eastern Idaho into southwestern Montana. Farther north, the province boundary is masked by younger thermal and structural events and by younger sedimentary and metasedimentary cover in Canada. The northern edge of the province coincides with a geophysical anomaly in Alberta that has been interpreted as a possible suture between the Wyoming Province and the Hearne Province of Canada (Hoffman, 1988; Houston, 1993).

Near the close of the Archean, a thick, cool, lithospheric mantle "keel" existed beneath the Wyoming Province. Such lithospheric roots are thought necessary to provide favorable conditions to produce magmas from depths within the diamond stability field. These conditions continued into Cretaceous time; however, Eggler and others (1988) and Eggler and Furlong (1991) suggested that portions of the lithosphere eastward towards the Great Plains were destroyed by Cretaceous to Eocene tectonomagmatism, such that remnants of the "keel" are interpreted to exist only in central Montana, southeastern Wyoming, and the Leucite Hills of southwestern Wyoming. Nevertheless, based on the presence of numerous and widespread kimberlitic indicator minerals found in Wyoming, a substantial mantle lithosphere may have existed under much of Wyoming during the geologic past.

The Colorado Province to the south contains the largest known field of kimberlites in the United States. This province consists of a 1.8 to 1.7 Ga metamorphic terrane disconcordantly intruded by 1.5 to 1.4 Ga granitic rocks (Peterman and others, 1968). The metamorphic rocks are amphibolitic, quartzofeldspathic, and pelitic gneisses and schists that are part of an island-arc succession that collided with the Wyoming Province 1.7 billion years ago. The collision boundary, marked by the Cheyenne belt, is interpreted to intersect the Laramie Mountains some 50 to 70 miles north of the State Line district (McCallum and others, 1977; Karlstrom and Houston, 1984).

Much of the lithosphere under the Colorado Province has thinned since Cretaceous time. However, a thick "keel" extended under the Colorado-Wyoming State Line district during the Paleozoic as is evidenced by the presence of several diamond-bearing kimberlites (Eggler and others, 1988). Eggler and others (1988) and Shaver (1988) proposed that a fragment of an Archean "keel" extended under the Proterozoic basement, although samples from State Line kimberlites suggested to Coopersmith and Schulze (1996) the upper mantle at depth contains only 5 to 10% harzburgite, which is more typical of a Proterozoic lithosphere.

Most kimberlites in the Colorado Province are early Devonian in age based on the presence of Paleozoic xenoliths, fission track dating of sphene (Larson and Amini, 1981) and zircon (Naeser and McCallum, 1977), and on Rb-Sr dating of phlogopite (C.B. Smith, 1979, 1983). However, some recent dates obtained from a few kimberlites suggest at least two episodes of emplacement may have occurred: a Late Proterozoic event (600 to 700 Ma) and a Devonian episode (A.P. Lester, personal communication, 1996). In addition to these events, other episodes of mantle-derived magmatism have been recognized in the Wyoming craton (Hausel, 1996a).

The majority of the known kimberlites occur within two districts: the State Line and Iron Mountain districts. A few scattered intrusives have also been found in the Estes Park region and the Boulder area of Colorado; the Sybille Canyon region of Wyoming; and the Missouri Breaks region of Montana. Some of the Wyoming kimberlites lie adjacent to terrane of several hundred square miles in extent that has produced numerous kimberlitic indicator mineral anomalies in stream sediment samples, suggesting that the known kimberlite fields are more widespread. Additionally, similar indicator mineral anomalies have also been identified along the flank of the Sierra Madre, Seminoe, and Bighorn Mountains; in the Medicine Bow and Uinta Mountains; and in the Green River Basin. To date, diamonds have been reported in kimberlites in the Colorado-Wyoming State Line district and in breccia pipes in the Green River Basin. Diamonds have also been found in the Medicine Bow Mountains and the Powder River Basin. There are also reports of diamonds from several other localities in Wyoming, Montana, and Idaho (Hausel, 1995a).

In addition to kimberlites, the Wyoming craton also encloses the largest lamproite field in the United States, known as the Leucite Hills. The Leucite Hills in southwestern Wyoming lie northeast of a highly anomalous 500- to 1000-mi^2 region containing widespread kimberlitic indicator minerals found in anthills, soils, stream sediments, and in the Bishop Conglomerate (Oligocene) (McCandless, 1982, 1984). This anomaly extends into extreme northwestern Colorado and northern Utah (McCandless and others, 1995). Other lamproites and some ultramafic lamprophyres have also been identified in the craton in Montana and northern Utah.

In southeastern Wyoming and northern Colorado, more than 100 late Proterozoic and early Devonian kimberlite intrusives and dozens of unexplored geophysical, remote sensing, and heavy mineral anomalies form the Colorado-Wyoming kimberlite province (Hausel, 1996a). Within the central portion of this province, 35 to 40 kimberlite dikes and diatremes (most of which are diamondiferous) occur in the State Line district (McCallum, 1991) (**Figure 7**). The district extends 4 miles north into Albany County, Wyoming, and 12 miles south into Larimer County, Colorado.

The known kimberlite intrusives extend many miles north and south of the Colorado-Wyoming State Line district. For instance, kimberlites have been found 50 to 60 miles south at Boulder, Colorado, and 40 to 45 miles north at Middle Sybille Creek in Wyoming. To date, only the State Line kimberlites have yielded diamonds, although subcalcic G10 pyrope garnets from the Iron Mountain district suggest those kimberlites also tapped the diamond stability field (McCallum and Waldman, 1991).

State Line district, Colorado-Wyoming

The Colorado-Wyoming State Line district extends from the southern Laramie Mountains of Wyoming into the Front Range of Colorado. Geographically, these two form a single, continuous, mountain range uplifted during the Laramide orogeny. Near the State Line district, the exposed Precambrian metamorphic crystalline rocks (1.9 to 1.7 Ga) of the Colorado Province are disconcordantly intruded by 1.4-Ga granitic rocks.

The dominant structure in the State Line district is the Virginia Dale ring-dike complex, which forms a circular granitic pluton nearly 9 miles in diameter. The ring-dike, interpreted as a phase of the Log Cabin batholith, intrudes the 1.4-Ga Sherman Granite and some metamorphic rocks (Eggler, 1967).

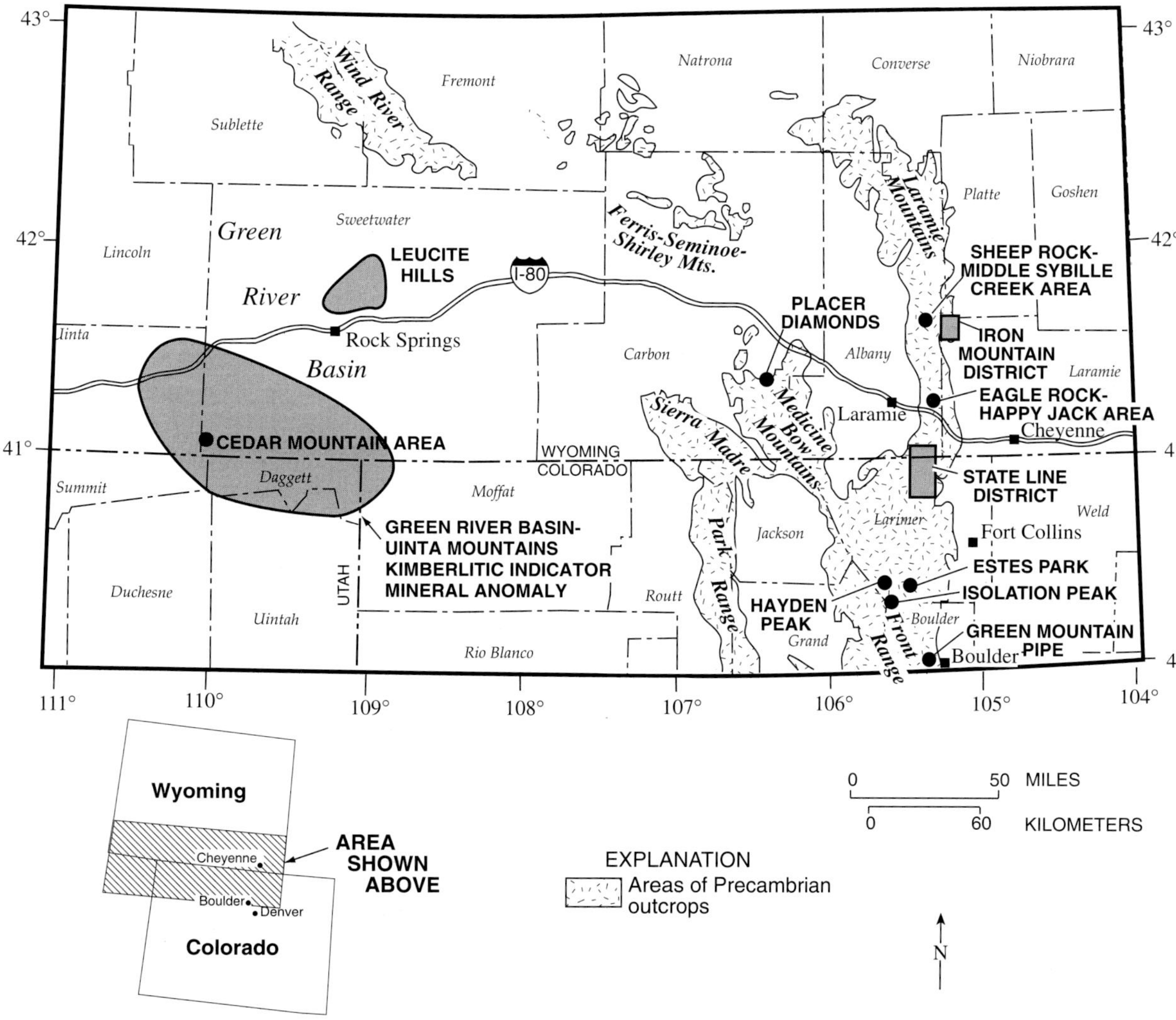

Figure 7. Location map of the Colorado-Wyoming kimberlite province and adjacent anomalous areas (modified from Hausel and others, 1985; Shaver, 1994).

The only post-Precambrian rocks found in the vicinity are arkosic sedimentary rocks of the Pennsylvanian Fountain Formation, located along the margins of the district, and Cambrian, Ordovician, and Silurian sedimentary xenoliths found in some kimberlite diatremes (McCallum and others, 1979). The kimberlites consist predominantly of hypabyssal and diatreme facies, although some minor crater facies kimberlite has been reported at the Kelsey Lake intrusives along the state line. Erosional models suggest that many of the known kimberlites may be deeply eroded with as much as 50% of the original pipes having been removed. Thus, many placer diamonds should lie downstream.

The diatreme facies kimberlite consists of abundant subrounded to angular rock fragments with serpentinized macrocrysts of olivine and lesser enstatite, Cr-rich and Cr-poor pyropic garnet, diopside, picroilmenite, clinopyroxene-ilmenite intergrowths, and phlogopite in a finely crystalline matrix of serpentine, carbonate, olivine, diopside, picroilmenite, phlogopite, perovskite, magnetite, Cr-rich spinels, hematite, apatite, and zircon (Rogers, 1985; McCallum, 1991). The hypabyssal facies kimberlite is a massive porphyritic rock that occurs in dikes, sills, and small plug-like bodies that represent root zones of deeply eroded pipes. The mineralogy of the hypabyssal kimberlite is essentially the same as the diatreme facies; however, globular, emulsion-like segregations of serpentine and calcite are more abundant in the hypabyssal facies (Rogers, 1985). Xenolithic fragments rarely exceed a few percent in the hypabyssal facies (unlike the diatreme facies), and typical fragments are moderately well-rounded upper mantle and lower crustal nodules.

Generally, the kimberlites are poorly exposed, have negligible relief, and are deeply weathered. Geophysical surveys suggest weathering continues to depths of at least 100 feet below the surface in many kimberlites, although moderately fresh exposures are present at some sites. Most kimberlites are covered by alluvium and/or colluvium, and some have been recognized by the presence of gray, weathered, "blue-ground" kimberlitic soils that are commonly associated with noticeable grassy vegetation anomalies (**Plates 2a** and **2b**; **Figures 8a** and **8b**). Other sites are recognized by the presence of Lower Paleozoic xenoliths as well as mantle and lower crustal xenoliths and nodules of peridotite, pyroxenite, eclogite, granulite, and/or megacrysts of pyrope, ilmenite, or diopside (McCallum and Eggler, 1979; Hausel and others, 1979).

Kimberlite was first recognized in the district in 1964 by M.E. McCallum from samples collected at the Sloan 1 diatreme. Many subsequent discoveries were made by D.H. Eggler, M.E. McCallum, and many of McCallum's graduate students at Colorado State University. Diamonds were first identified in the district in 1975 during routine thin-section prepa-

a.

b.

Figure 8. The vegetation anomaly (lack of trees and higher stand of grass) is noticeable over some kimberlite intrusives. (a) View looking along contact of the Schaffer 15 kimberlite with the country rock, and (b) view looking down along contact of a kimberlite at Iron Mountain with the country rock granite. In both areas the kimberlites lie under the high grass on the right side of each photograph.

ration of a serpentinized garnet peridotite nodule collected from the Schaffer 3 kimberlite in Wyoming. During the preparation of the section, deep scratches were cut into a carborundum grinding plate by a small, white, crystal that was later confirmed as diamond (McCallum and Mabarak, 1976).

To date, total diamond production from the district has amounted to more than 130,000 gem and industrial stones (**Table 2**). The diamonds range from microdiamonds to a 28.3-carat octahedron, which is the fifth largest verified diamond found in the United States (**Table 1**). The gem to industrial quality diamond ratios in the district are very favorable, with some deposits reporting 65% gemstones.

The district's diamond population is dominated by colorless stones, although black, gray, brown, yellow, and lesser green and pink diamonds have also been reported. The colorless stones are in many cases high-quality gemstones. Information on the pink diamonds is lacking, which is intriguing because other pink diamonds, known as the "Argyle pinks" from Australia, represent some of the most valuable gemstones available on the market.

Additional exploration in the district will undoubtedly lead to the discovery of more kimberlites. For example, an airborne INPUT® survey over the Wyoming portion of the district in 1981 identified several conductivity anomalies that were not associated with any known kimberlites. In addition, several weak to strong magnetic anomalies were detected. One group of prominent magnetic anomalies located along the northern edge of the district were interpreted as undiscovered (buried or blind) pipes (Paterson and MacFadyen, 1984). To the author's knowledge, as of 1997 none of these latter anomalies had been drilled!

Another study in the northern part of the district, identified several remote sensing anomalies possibly associated with kimberlite (Marrs and others, 1984; Marks, 1985). Similar studies in the district should result in the discovery of other hidden pipes. And since all the kimberlites tested in the district to date have been diamondiferous, one would anticipate to see additional INPUT® and remote sensing studies in the future.

Approximately one-half of the known kimberlites in the State Line district occur in the Colorado portion of the district. Diamonds have been recovered from most of these kimberlites, and some gold prospectors from Fort Collins have also reported finding diamonds in the Poudre River, which drains portions of the district. Other placer diamonds were mentioned in the San Francisco Chronicle (August 1, 1872), which stated that the discovery consisted of several small diamonds of no commercial value. The article indicated the diamonds were found by a group of gold prospectors on the "headquarters" [headwaters?] of the Santa Maris, a branch of the Bill Williams Fork of the Colorado River. Since this news article was published in 1872, it is suggested that there may be a possible connection to the "Great Diamond Hoax of 1872" (e.g., Hausel and Stahl, 1995).

Schaffer, Ferris, and Aultman kimberlites

The Colorado-Wyoming State Line district is located 20 miles south of Laramie, and immediately west of U.S. Highway 287 (**Figure 9**). On the Wyoming side of the district, 20 kimberlite intrusives known as the Schaffer group (complex), Aultman 1

Table 2. Reported diamond recoveries from State Line district kimberlites exclusive of Kelsey Lake (McCallum and Waldman, 1991; Shaver, 1994).

Site	Tonnes processed	Total stones	Average grade carats/100 tonnes	Range of grades carats/100 tonnes
Maxwell	300	28	<0.5	?
Aultman	1859.5	132	<1.0	?
Schaffer group	3892.6	227	<1.0	?
Chicken Park	295.9	306	6.7	?
Sloan 1	1830.2	13,749	8.7	1.2-21.3
Sloan 2	4107.6	25,867	12.68-18.4	2.2-59.8
Sloan 5	903.5	474	1.0	0.0-5.3
Sloan 6	500.0	215	1.3	0.2-4.8
George Creek K1	2613.2	80,282	46.1	20.0-135.1
George Creek K2	416.1	8873	31.0	18.1-86.8
Total	**16,718.6**	**130,153**		

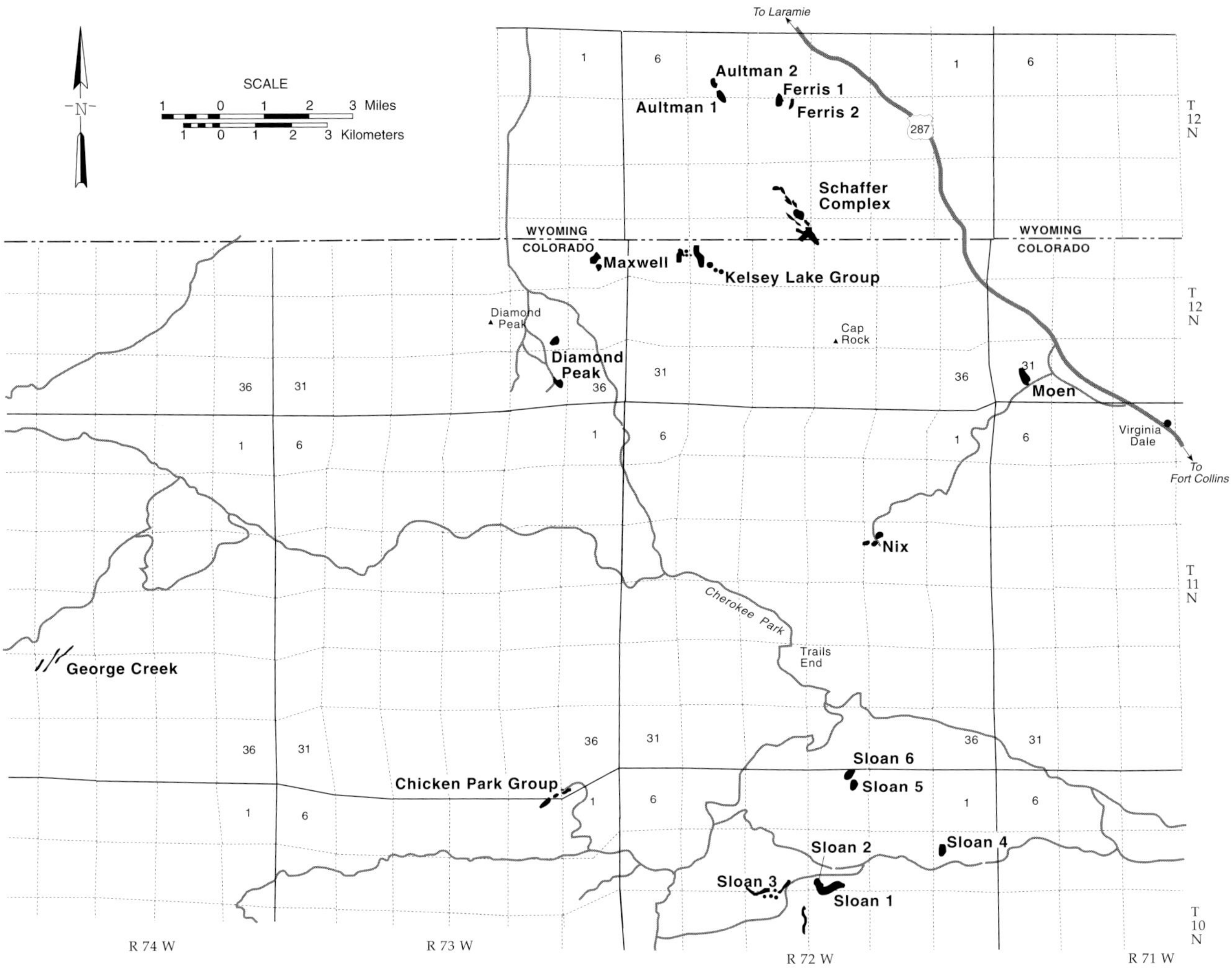

Figure 9. Location map of the known kimberlites in the Colorado-Wyoming State Line district.

and 2, and the Ferris 1 and 2 lie within 2 to 3 miles of the Kelsey Lake diamond mine (**Figure 9**). These occur as dikes, blows, and diatremes and contain hypabyssal and diatreme facies kimberlite (McCallum and Mabarak, 1976; Hausel and others, 1981).

Diamonds recovered from bulk samples collected from the Schaffer and Aultman kimberlites by Cominco American Incorporated in the early 1980s (**Plates 2c, 2d,** and **2e**) included approximately 50% stones of gem quality (**Figure 10**). The largest stones weighed 0.985 and 0.86 carats, respectively, and were recovered from the southern part of the Schaffer complex. The average grades of the bulk samples were low, ranging from 0.5 to 1.0 carat/100 tonnes (Hausel, 1993b). However, less than 9000 tonnes of material were mined from a group of about ten kimberlites in the district (typically, 10,000 tonnes of kimberlite are sampled per each intrusive during feasibility studies), and some kimberlites still remain untested to date.

In addition to the known kimberlites, several geophysical anomalies and anomalous stream sediment samples suggest the presence of additional undiscovered kimberlites within the district. In a

Figure 10. Industrial and gem-quality diamonds recovered from Wyoming State Line kimberlites in the early 1980s. The six diamonds on the right side of the photograph are gem quality.

Plate 2

Plate 2a. Blue ground (weathered kimberlite) exposed in badger diggings on the Schaffer 15 kimberlite, Wyoming.

Plate 2b. The different soil colors are readily visible in drill hole cuttings between the contact of the Aultman 1 kimberlite with the adjacent Sherman Granite. The blue ground on the right side of the photograph is distinctly different from the granitic soils on the left.

Plate 2c. Bulk sampling on the Aultman 1 kimberlite, Wyoming, by Cominco American Incorporated. Trenching of the kimberlite exposed the characteristic blue ground.

Plate 2d. Trenching in the Schaffer kimberlites farther to the south of Plate 2c revealed a sharp contact with the host granite.

Plate 2e. Close inspection of the kimberlite/granite contact at the Schaffer kimberlites, Wyoming, showed a knife-sharp contact with the granite and no evidence of wall rock alteration typical of kimberlites worldwide. The cold emplacement temperatures of the kimberlite magma are due to adiabatic expansion of CO_2 gas from the magma, which freezes the contact. Kimberlite magma is interpreted to be emplaced at the surface at a temperature of about 0°C.

news release of September 19, 1996, Royal Gold reported finding indicator minerals in the district: "... in drainages where kimberlites were not previously known to exist." Of 73 samples collected by the company, indicator minerals were found in 17 samples. In an earlier survey, anomalous samples were also recovered immediately north of the known kimberlites (Hausel and others, 1988).

Geophysical studies have also provided evidence of buried or blind diatremes in the district. During ground geophysical surveys, Hausel and others (1979, 1981) identified electrical conductors and weak to moderate magnetic anomalies associated with several known kimberlites. This led to the identification of a kimberlitic blow buried in a drainage along the trend of the Schaffer kimberlites in Wyoming. A later airborne INPUT® survey over the district not only detected most of the known kimberlites, but also identified several possible blind diatremes (**Figure 11**). Paterson and MacFadyen (1984) reported that INPUT® responses corresponded well with the known kimberlite intrusives in the district. Conductors were detected in association with the Schaffer group of kimberlites in both Colorado and Wyoming (the Schaffer kimberlites in Colorado later became part of the Kelsey Lake diamond mine). Additionally, a group of 10 to 12 fairly strong, discrete, magnetic anomalies identified by Paterson and MacFadyen's (1984) survey in the vicinity of the Ferris pipes were interpreted to be manifestations of buried, near-surface diatremes. One of these magnetic anomalies was also located over the Kelsey Lake area.

The presence of these untested geophysical anomalies and kimberlites within 3 miles of the Kelsey Lake diamond mine in a known diamond district is astounding! To date, there has been no research on the potential for placer diamonds in this area, yet many of the kimberlites in the district appear to be deeply eroded.

Several kimberlites occur in the Colorado portion of the State Line district in Larimer County (Braddock and others, 1989; Coopersmith, 1991; McCallum, 1991). These include the Kelsey Lake kimberlites; the Chicken Park kimberlites; the Sloan 1, 2, 3, 4, 5, and 6 kimberlites; the Moen intrusive (located in section 31, T12N, R71W); the Nix 1, 2, 3, and 4; the Maxwell pipes; and the Diamond Peak kimberlites (**Figure 9**) (Shaver, 1994). Other kimberlites in the district include the George Creek dikes and the Pearl Creek intrusive located south of George Creek (Shaver, 1994).

Kelsey Lake kimberlites

These kimberlites are located in sections 19 and 20, T12N, R72W. Commercial production of diamonds began in 1996 in the Colorado-Wyoming State Line district, after Redaurum Ltd. placed two of the Kelsey Lake kimberlites (KL1 and KL2) into production (**Plates 3a** and **3b** and **Figure 12**). These lie near the Wyoming border in Colorado. A group of kimberlites were discovered in this area by Eggler (1967) and named the Schaffer 1, 2, 6, 7, 8, and 9 (see also McCallum and others, 1975; McCallum and Mabarak, 1976). Years later, some additional kimberlites were discovered by Coopersmith (1991, 1997), and the new discoveries along with the earlier Schaffer kimberlites were designated as the Kelsey Lake pipes.

The Kelsey Lake kimberlites yielded magnetic and conductivity anomalies that were detected by some early remote sensing and INPUT® surveys (**Figure 11**) (Marrs and others, 1984; Paterson and MacFadyen, 1984).

The Kelsey Lake kimberlites are irregular-shaped pipes and fissures containing diatreme facies kimberlite with zones of hypabyssal facies and minor crater facies kimberlite. An apparent Devonian age on the Kelsey Lake kimberlites is in agreement with Early Devonian isotopic ages for most other pipes found in the Colorado-Wyoming kimberlite province (Coopersmith, 1993, 1997).

To date, these kimberlites and adjacent alluvial material have yielded many high-quality diamonds heavier than 1 carat (**Figure 13**). Some of the larger recovered stones have included 6.2-, 9.4-, 10.48-, 11.85-, 14.2-, 16.9-, 28.18-, and 28.3-carat gemstones. The diamonds have predominantly octahedral habit, and are colorless with some honey-brown gemstones (Coopersmith and Schulze, 1996). The probability that many other large gemstones will be found in the future is considered high, as the property has only been in production for a short time.

One of the largest stones recovered from the pipe, a 28.18-carat diamond, was cut and produced the largest cut U.S. diamond. The cut diamond weighed 16.8 carats, and had an estimated value of more than $250,000 (Denver Post, September 25, 1997). A 28.3-carat diamond, also recovered from Kelsey Lake, was cut into a 5.39-carat gemstone that sold for $87,000 (Paydirt, 1996).

The two largest of the Kelsey Lake pipes cover nearly 20 surface acres. A mill on the property is

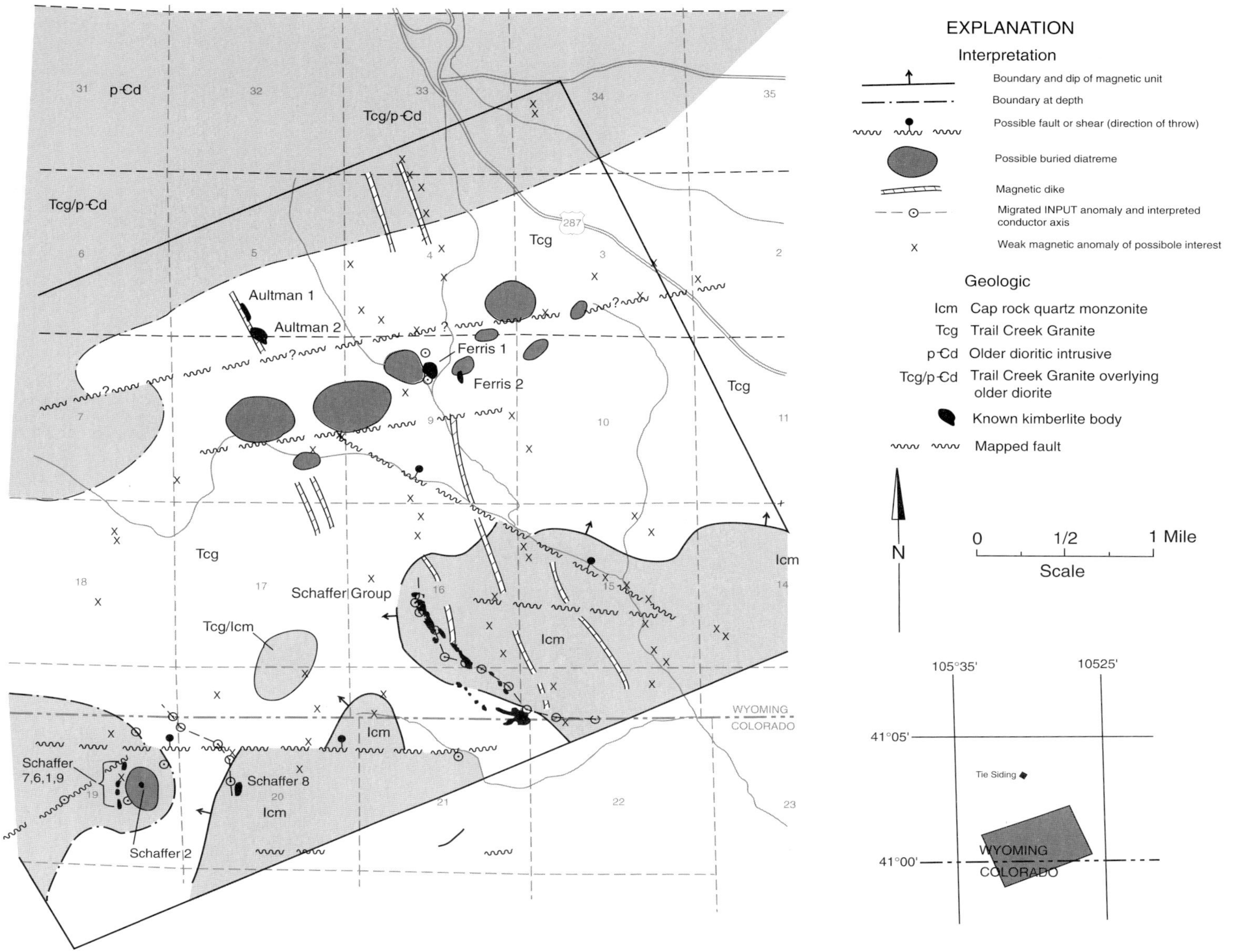

Figure 11. Airborne INPUT® survey of a portion of the Colorado-Wyoming State Line district showing magnetic and conductivity anomalies (modified from Paterson and MacFadyen, 1984).

Figure 12. The kimberlite at Kelsey Lake is weathered and very soft; as a result, the material is scraped and no blasting is necessary.

Figure 13. A sampling of several gem-quality diamonds recovered from the Kelsey Lake diamond mine. The largest stone in the center weighed 14.2 carats. Photograph courtesy of Redaurum Ltd. and Howard Coopersmith.

designed to operate at 195 tonnes/hour and process approximately 40,000 to 50,000 tonnes/month (**Figure 14**) (Coopersmith and Schulze, 1996). The mine is expected to have at least a 12-year life, and resources have been established at 16.9 million short tons to a depth of 320 feet (Coopersmith, 1997).

Indicator minerals from the Kelsey Lake kimberlites include abundant G9 pyrope garnets, lessor sub-calcic G10 Cr-pyropes, and high-Cr chromites. The ilmenite compositions show magnesium depletion and chrome enrichment similar to ilmenites from the Iron Mountain, Wyoming kimberlites. (Ilmenite compositions are thought to provide information on the reducing/oxidizing conditions of the kimberlite magma, and thus provide an indication of diamond preservation.) The compositions of the Kelsey Lake ilmenites suggest that the kimberlite magmas were oxidizing, implying that the diamonds should have been badly resorbed and

Figure 14. The Kelsey Lake diamond mill with piles of kimberlite in foreground.

destroyed during emplacement. However, the presence of gem-quality diamonds with octahedral habit suggests that the ilmenite compositions are not providing a reliable barometer on diamond preservation at Kelsey Lake (Coopersmith and Schulze, 1996).

Chicken Park kimberlites

These kimberlites are located in sections 35 and 36, T11N, R73W, and section 2, T10N, R73W. The Chicken Park kimberlites include three relatively small blows of lower diatreme facies and upper hypabyssal facies kimberlite, and one dike of hypabyssal facies kimberlite that all lie along a northeasterly trend. The largest intrusive is a blow of about 250 by 160 feet. Three of the intrusives were outlined by vegetation anomalies; the fourth was not exposed at the surface and was later discovered following stream sediment sampling surveys (Rogers, 1985).

The stream sediment sampling surveys were completed by Rogers (1985) in the Red Feather area and included about 200 samples. Of these samples, 12 contained picroilmenite, and a few contained both picroilmenite and pyrope garnet. Some of the anomalies were produced from material derived from the exposed Chicken Park kimberlites as the anomalies lie immediately downstream from the intrusives, and a few anomalies were derived from the buried CK1 intrusive. Three anomalous samples were taken in ephemeral drainages upstream from the Chicken Park intrusives, suggesting the presence of another buried intrusive, or intrusives, in the Chicken Park area.

Bulk sample tests on the intrusives resulted in the recovery of 306 diamonds from a 296-tonne sample at an average grade of 6.7 carats/100 tonnes. The largest recovered diamond was a 2.6-carat industrial stone.

The geochemistry of indicator minerals shows that the ilmenites have intermediate to high MgO content and are depleted in Cr_2O_3, suggestive of oxidizing conditions. Garnets from the property have yielded sub-calcic, high-Cr, harzburgitic pyropes (G10), along with a large population of calcic G9 pyropes (McCallum and Waldman, 1991).

The age of the kimberlites appears to be Late Proterozoic, even though most kimberlites in the Colorado-Wyoming State Line district are Early Devonian. Ar^{40}-Ar^{39} and Rb-Sr data from a sample of the Chicken Park kimberlites yielded Late Proterozoic ages similar to the George Creek dikes and Green Mountain pipe (A.P. Lester, personal communication, 1996).

Sloan kimberlites

Six kimberlites in the southern portion of the State Line district (**Figure 9**) in the Rabbit Creek area include the Sloan 1 and 2 (located in section 10, T10N, R72W); Sloan 3 (sections 9 and 16, T10N, R72W); Sloan 4 (section 12, T10N, R72W); and Sloan 5 (**Figure 15**) and 6 (section 3, T10N, R72W). Four of the six kimberlites have been bulk sampled (**Table 2**). There is no information available on the Sloan 3 and 4 kimberlites.

Figure 15. Aerial view of the Sloan 5 kimberlite. This kimberlite lies in a distinct open park and along a prominent structure highlighted by a linear stand of trees in the bottom right corner of the photograph.

The Sloan 1 and 2 kimberlites consist of hypabyssal and diatreme facies kimberlite. The Sloan 1 intrusive has an areal extent of 500 by 1800 feet and the Sloan 2 is 200 by 2000 feet. The Sloan 1 kimberlite was discovered in 1964 by M.E. McCallum based on exposures of kimberlite along its western edge, whereas the Sloan 2 kimberlite was discovered several years later by C.D. Mabarak following a stream sediment sampling survey in the area (McCallum and Mabarak, 1976). The Sloan 1 and 2 have yielded a significant number of sub-calcic G10 harzburgitic pyropes, indicating a moderate potential for peridotitic diamonds; the TiO_2/Na_2O contents of the eclogitic garnets indicate a low potential for eclogitic diamonds (McCallum and Waldman, 1991). Despite this, the majority of the diamonds recovered from the Sloan 1 and 2 are E-type diamonds (Otter and Gurney, 1989).

The $MgO:Cr_2O_3$ ratios in picroilmenites from Sloan 1 and 2 suggest that the magma was reducing and favorable for diamond preservation (McCallum and Waldman, 1991). The recovered diamonds range from well-preserved octahedra to highly resorbed tetrahexahedra with colors ranging from D to E on the GIA (Gemological Institute of America) color grading system (see Bruton, 1979, p. 279) and include stones that are very dark yellow-brown (Dummett and others, 1988).

The Sloan 1 intrusive is estimated to host a resource of 15.3 million tonnes of ore at an average grade of 6.1 carats/100 tonnes. The Sloan 2 intrusive is estimated to contain a resource of 8.4 million tonnes at an average grade of 17.1 carats/100 tonnes. Production from the Sloan 1 and 2 amounted to 21,546 diamonds in the 1980s (Oliver, 1990); more recent bulk sampling by Royalstar Resources on the Sloan 2 kimberlite yielded 9034 diamonds larger than 1 mm diameter with a combined weight of 342.17 carats from 3300 tonnes of rock (**Figure 16**). The average grade of the Royalstar sample was 12.68 carats/100 tonnes (Shaver, 1994).

a.

b.

c.

Figure 16. (a) Aerial view of the Superior diamond mill and ore piles in 1982; (b) view of the Royalstar adit driven into the Sloan 2 kimberlite with Robert Gregory in foreground; and (c) a large granite xenolith trapped in the kimberlite and exposed in the mine rib next to Bernie Free.

The largest diamond recovered from the Sloan 1 and 2 was a 5.51-carat gemstone (**Figure 17**) (Bernie Free, personal communication, 1994), which was expected to produce a clear, white, brilliant-cut stone weighing at least 2 carats (Shaver, 1994). Cumulative production from the Sloan 1 and 2 kimberlites has totaled 39,616 diamonds (modified from Hausel, 1996b).

The Sloan 5 intrusive lies to the north of the Sloan 1 and 2, and was bulk sampled in 1982. In total, 474 diamonds totaling 9.09 carats were recovered from 904 tonnes (an average grade of 1.0 carat/100 tonnes). Bulk sampling of the nearby Sloan 6 intrusive resulted in the recovery of 215 diamonds (total weight of 6.72 carats) from 500 tonnes at a grade of 1.3 carats/100 tonnes.

Nix kimberlites

A group of four intrusives located in section 14, T11N, R72W, form the Nix cluster of kimberlites. These include the NX1 and NX3 kimberlites, which are formed primarily of brecciated diatreme facies kimberlite, and the NX2 and NX4 kimberlites, which consist of massive hypabyssal facies kimberlite porphyry. A 300-pound sample collected from two of the four Nix kimberlites yielded microdiamonds (McCallum and others, 1977).

Maxwell pipes

Two kimberlites, known as the Maxwell pipes, lie near the Wyoming border in section 24, T12N, R73W. The Maxwell 1 forms a topographic depression marked at the surface by limestone nodules and a typical suite of kimberlitic indicator minerals in the soil over the pipe (**Figure 18**). The intrusive covers a surface area of about 5 acres. The Maxwell 2 also forms a topographic depression and is marked by a pronounced vegetation anomaly (Carlson, 1983).

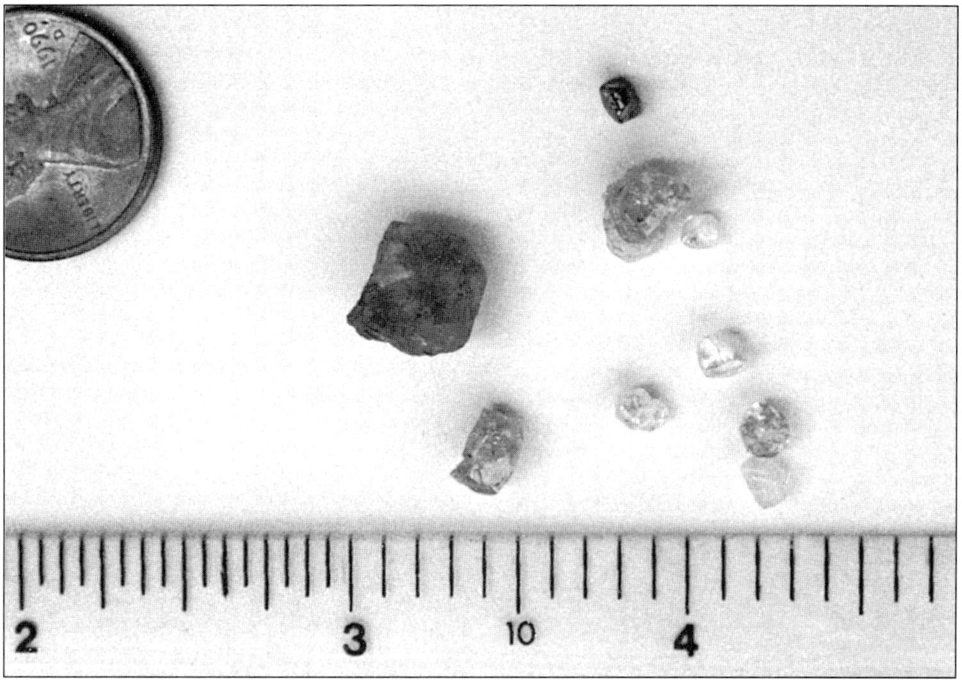

Figure 17. Diamonds from the Sloan 2 kimberlite, including a 5.51-carat stone. Scale is in inches.

Diamond Peak kimberlites

These kimberlites are located in sections 26 and 35, T12N, R73W. Two kimberlites, known as the Diamond Peak kimberlites, were identified following stream sediment sampling surveys near Diamond Peak. According to Carlson (1983), the DP1 kimberlite occurs in a small alluvial valley and is marked by limestone xenoliths and weathered kimberlite in animal burrows. The DP2 kimberlite is a small intrusive (< 200 feet in diameter) marked by a topographic depression with an overlying vegetation anomaly.

Figure 18. The Maxwell 1 kimberlite is expressed as a topographic depression with a grassy vegetation anomaly shown in the center of the photograph.

George Creek kimberlites

The George Creek intrusives, located about 15 miles southwest of Kelsey Lake in sections 21 and 29, T11N, R74W, were discovered following regional stream sediment sampling surveys that detected kimberlitic indicator minerals in the nearby drainages. The pyrope populations included several subcalcic, high-Cr, harzburgitic G10 pyropes.

The source of the indicator minerals was identified after follow-up VLF (very low frequency) electromagnetic surveys detected a northeasterly trending buried conductor upstream from the anomalous sediment samples. This conductor was later trenched exposing kimberlite (Carlson and Marsh, 1989).

Three narrow, steeply dipping kimberlite dikes that are known as the K1, K2, and K3 intrusions were found. These dikes range in length from several hundred to about 3000 feet, with widths of inches to 13 feet. The dikes consist of hypabyssal macrocrystalline phlogopitic kimberlite. Rb-Sr analysis of weathered, surficial material from one of the dikes yielded a preliminary age of about 600 Ma (Carlson and Marsh, 1989) similar to the Chicken Park and Green Mountain kimberlites.

The dikes were relatively diamond rich and yielded more than 89,000 diamonds during testing. Recovered bulk samples from individual test pits yielded grades ranging from 18 to 135 carats/100 tonnes, whereas average grades for the K1 and K2 dikes at George Creek yielded 46.1 and 31.0 carats/100 tonnes, respectively (McCallum and Waldman, 1991).

Diamonds from George Creek were apparently emplaced under reducing magmatic conditions, as the diamonds show a low degree of resorption, and include octahedra, macles, aggregates, and dodecahedra (tetrahexahedra). The colors include a high proportion of colorless stones with some yellow, brown, green, and gray diamonds (Falk, 1992).

Kimberlites and related rocks and minerals in Colorado outside the State Line district

Baca County

The Two Buttes igneous complex in southeastern Colorado includes laccoliths, sills, and dikes of monazite to minette affinity. The minettes are olivine lamprophyres rich in iron, magnesium, and alkalies (particularly potassium) (Kreiger and Thornton, 1976).

Boulder County

A single, 128-foot diameter diatreme, known as the Green Mountain pipe, intrudes Boulder Granite in the Boulder City Park along the edge of the Front Range in north-central Colorado, 50 to 60 miles south of the State Line district (**Figure 7**) (Meyer and Kridelbaugh, 1977; Hausel and others, 1985). Similar material 6 to 8 miles west of the diatreme was reportedly identified at the end of the 19th century (Whitaker, 1898), although no other kimberlite is presently known in this region. The pipe yielded paleomagnetic, K-Ar, Ar^{40}-Ar^{39}, and Rb-Sr ages between 600 to 700 Ma (A.P. Lester, personal communication, 1996). No diamonds have been reported from the pipe.

Custer County

Peridotite was reported near Querida in Custer County of southern Colorado. The peridotite forms a small, oblong outcrop on the north bank of Cottonwood Gulch above the Mountain Boy mine. The rock consists of hornblende, hypersthene, and olivine (Blank, 1934, 1935). No kimberlites or lamproites are reported in the county.

Larimer County

Other kimberlites occur in Larimer County outside the State Line district but within the Colorado-Wyoming kimberlite province (**Figure 7**). These include a single, 3- to 6.5-foot wide kimberlite dike 38 miles south of the State Line district and south of Estes Park near Ramshorn Mountain (sections 1 and 12, T4N, R73W). Preliminary testing of the dike yielded no diamonds (Hausel and others, 1985). Other kimberlite dikes are reported west of Estes Park in Hayden Gorge adjacent to Hayden Spire [Latitude (Lat.) 40°21'30"N, Longitude (Long.) 105°43'30"W], and southwest of Isolation Peak (Lat. 40°12'30"N, Long. 105°42'W) in Rocky Mountain National Park (Braddock and Cole, 1990).

Moffat County

Diamonds, rubies, sapphires, emeralds, and pyrope garnets all occur on a sandstone outcrop in Diamond Field Draw along the northeastern flank of Diamond Peak in the northwestern corner of Colorado, a short distance south of the Wyoming border (**Figure 19**). In 1986, the author recovered 4 diamonds, 17 rubies, and 24 pyrope garnets from this area. This was the site of an elaborate fraud known as the "Great Diamond Hoax" that took place in 1871 and 1872. Gemstones purchased in London and Arizona by two enterprising individuals were scattered on the surrounding sandstone outcrops as part of an elaborate fraud (Hausel, 1994a; Hausel and Stahl, 1995).

In one of the more extraordinary coincidences in history, the salted area overlies an Archean craton between the Colorado-Wyoming State Line district and the Cedar Mountain breccia pipes in the Green River Basin. Even more remarkable, similar kimberlitic indicator minerals–pyrope garnet, chromian diopside, and chromian enstatite–all occur in outcrops of Bishop Conglomerate at the top of Diamond Peak (Hausel and Stahl, 1995; McCandless and others, 1995). In an almost unbelievable coincidence, the garnets used in the salting were apparently derived from breccia pipes in the Navajo volcanic field (see Arizona, p. 48), and at the time of the fraud, few people, if any, were aware of the association of kimberlite, kimberlitic indicator minerals, and diamonds, nor was there any known association of diamonds with cratons.

Figure 19. Site of the "Great Diamond Hoax of 1872." The sieve piles on the sandstone outcrop are the historic remnants of sampling by the 40th parallel survey in 1872 to prove the site was a hoax. The diamonds, rubies, sapphires, and pyrope garnets found at the site were placed on sandstone and in nearby anthills in an attempt to fool investors.

Kimberlites, lamproites, and related minerals in Wyoming outside the State Line district

Albany County

Throughout a large portion of Albany County (and extending into adjacent Laramie County) in both the Laramie and Medicine Bow Mountains, stream sediment sampling surveys were conducted for several years by the Wyoming State Geological Survey in the search of kimberlite. The project was successful and resulted in the identification of approximately 300 kimberlitic indicator mineral anomalies (Hausel and others, 1988), but due to funding limitations, none of the minerals were geochemically analyzed.

These anomalies were identified in the Laramie Mountains north of the State Line district, and a few were identified in the Medicine Bow Mountains to the west. The results indicate the presence of several undiscovered kimberlite intrusives. The potential for diamond discoveries in these areas is considered high in that the geochemistry of all kimberlites examined to date in the Laramie Mountains suggest they originated within the diamond stability field.

Four areas in particular stand out as highly anomalous. These include the Eagle Rock (Happy Jack-Pole Mountain) and Middle Sybille Creek (Sheep Rock) areas, Grant Creek, and the Elmers Rock greenstone belt (**Figure 7**) (Hausel and others, 1997).

Eagle Rock area. In the Eagle Rock area, 35 stream sediment samples collected over 30 mi^2 yielded pyrope garnet, chromian diopside, and/or picroilmenite (**Figure 20**). Many of the samples lie immediately downstream from beaver ponds. During aerial photo reconnaissance, a prominent circular structure (the Eagle Rock anomaly) was identified along a north-northwesterly trending lineament (**Plate 3c**). Field reconnaissance of the anomaly identified a small topographic depression with an associated vegetation anomaly similar to several of the kimberlites in the State Line district (**Figure 21**). The anomaly is soil covered with no rocks exposed on the surface. One pyrope garnet was recovered from several pounds of soil collected during follow-up sampling within the structure. To date, the origin of the anomaly has not been determined.

Middle Sybille Creek area. Twenty to 25 miles north of Eagle Rock is the Middle Sybille Creek area near the central portion of the Laramie Mountains (**Figure 7**). A single blow known as the Radichal kimberlite, composed of hypabyssal facies kimberlite, was found in this area (Hausel and others, 1981). Stream sediment sampling nearby produced 32 anomalous samples which extend 5 miles upstream from the kimberlite (**Figure 22**), suggesting the presence of several undiscovered kimberlites in the area. Unverified reports have indicated that a few other kimberlites have recently been found in this area.

Grant Creek. Samples collected in Grant Creek northeast of Middle Sybille Creek were also anomalous. Twelve stream sediment samples collected over a 1-mile distance yielded several pyrope garnets, chromian diopsides, and/or picroilmenites (**Figure 23**). A short distance upstream, samples yielded no indicators. This evidence suggests the presence of a buried kimberlite nearby. In addition, pyrope

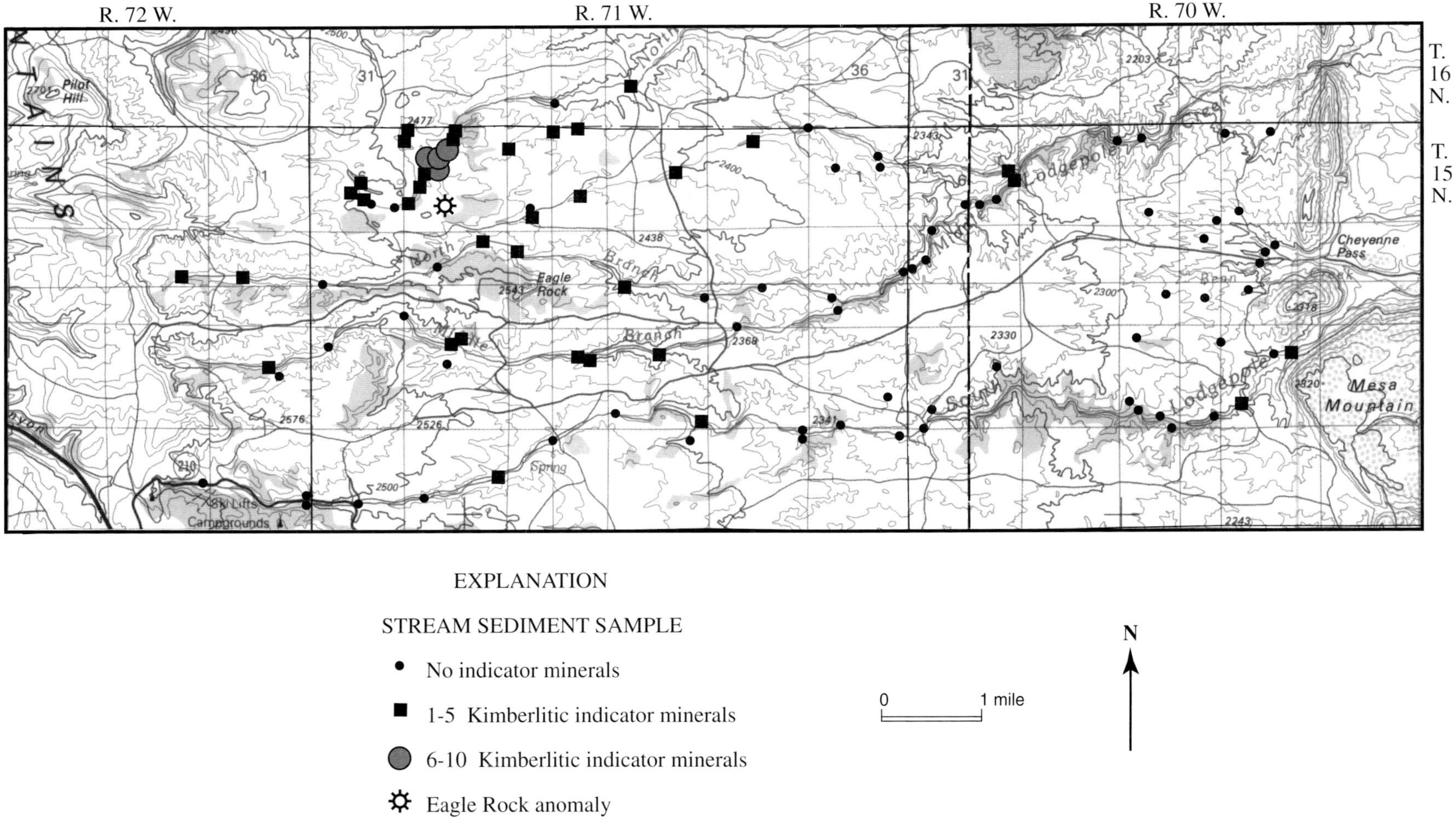

Figure 20. Map of the Eagle Rock area showing locations of sample sites and indicator mineral anomalies (after Hausel and others, 1988, 1997).

Figure 21. View of the Eagle Rock anomaly showing a distinct topographic depression filled with water and high grass, surrounded by aspen trees.

garnets were also recovered from samples provided to the Wyoming State Geological Survey by Bret Boundy from the Deadhead gravel pit in the immediate area.

Elmers Rock greenstone belt. Stream sediment samples collected in the Elmers Rock greenstone belt north of Grant and Middle Sybille Creeks also yielded dozens of anomalous samples (**Figure 24**). Some of the indicator minerals lie downstream from a 0.5-mile diameter circular structure of unknown origin. In addition, many other samples yielded several kimberlitic indicator minerals, suggesting the presence of nearby hidden kimberlites. A report of an unverified diamond from the Bluegrass Creek area of the Elmers Rock greenstone belt is also intriguing (Hausel, 1981).

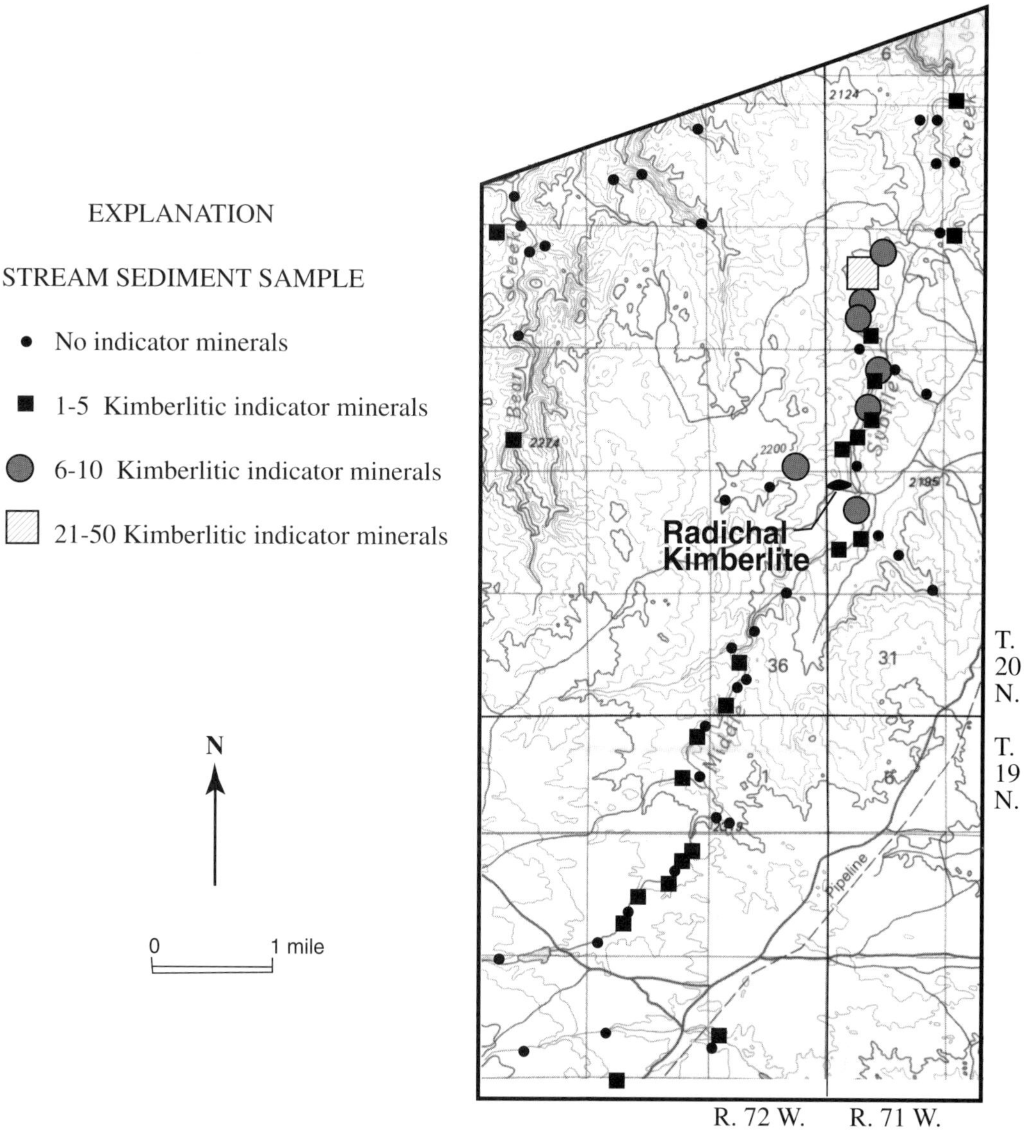

Figure 22. Map of the Middle Sybille Creek (Sheep Rock) area showing locations of anomalous samples and the Radichal kimberlite (after Hausel and others, 1988, 1997).

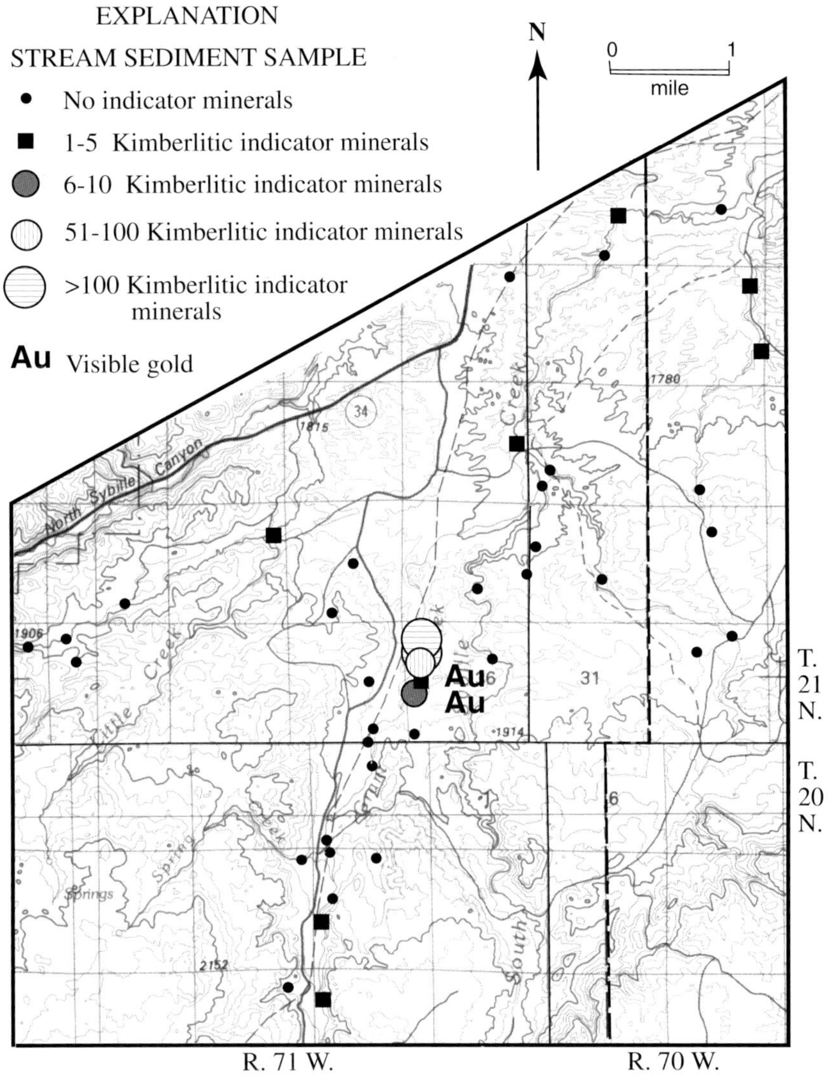

Figure 23. Map of the Grant Creek area showing locations of anomalous samples (after Hausel and others, 1988, 1997).

West of the Laramie Mountains and the Laramie Basin are the Medicine Bow Mountains, located in both Albany and Carbon Counties. The Medicine Bow Mountains have had a very limited amount of sampling, even though the range appears to be highly anomalous and has produced both indicator minerals and diamonds (see Carbon County, this page). Visible gold was also recovered in many of the samples in this region. This region should be considered high priority in any diamond exploration program.

Within the Medicine Bow Mountains, kimberlitic indicator minerals were recovered on Libby Creek, a short distance southwest of Barber Lake on the section line between sections 28 and 29, T16N, R78W; some indicator minerals were also recovered at the mouth of Camp Creek in the E/2 section 10, T15N, R79W (Hausel and others, 1988) (**Figure 25**). Pyrope garnets were also recovered in 1995 from a placer gold operation on Douglas Creek about 0.5 mile downstream from the Bobbie Thompson campground (Paul Allred, personal communication, 1995). These garnets were verified by petrographic analysis at the Wyoming State Geological Survey.

Campbell County

Microdiamonds were recovered from a coal seam near Gillette in the Powder River Basin, northeastern Wyoming (Finkelman and Brown, 1989). Some garnets were also recovered from the sample (R.B. Finkelman, personal communication, 1989). The source of the diamonds is unknown; but, with the recent discovery of Tertiary-age mantle-derived breccia pipes in the Green River Basin of southwestern Wyoming, exploration for similar pipes in the Powder River Basin should be considered. The Powder River Basin, like the Green River Basin, overlies the Wyoming Province (craton).

Carbon County

This area includes a portion of the Medicine Bow Mountains, the Sierra Madre to the west, the Seminoe Mountains to the north, and various Tertiary basins between them. Indicator minerals have been found associated with all three of these ranges, and diamonds have also been found in the Medicine Bow Mountains.

Cortez Creek. Two diamonds were recovered from a gold placer on Cortez Creek in the northern Medicine Bow Mountains, southeastern Wyoming, in 1977. The Boden diamonds weighed 0.1 and 0.03 carat (Hausel and others, 1985). Although no kimberlitic indicator minerals were found on Cortez Creek during follow-up sampling, indicator minerals were found in Iron Creek (a tributary of South French Creek) and in North French Creek a few miles south of Cortez Creek. The indicators were believed to have been derived from the Ferris-Hanna Formation (Tertiary). A third diamond was recovered from core drilled into Precambrian metaconglomerate in the northern Medicine Bow Mountains (T.E. McCandless, personal communication, 1991).

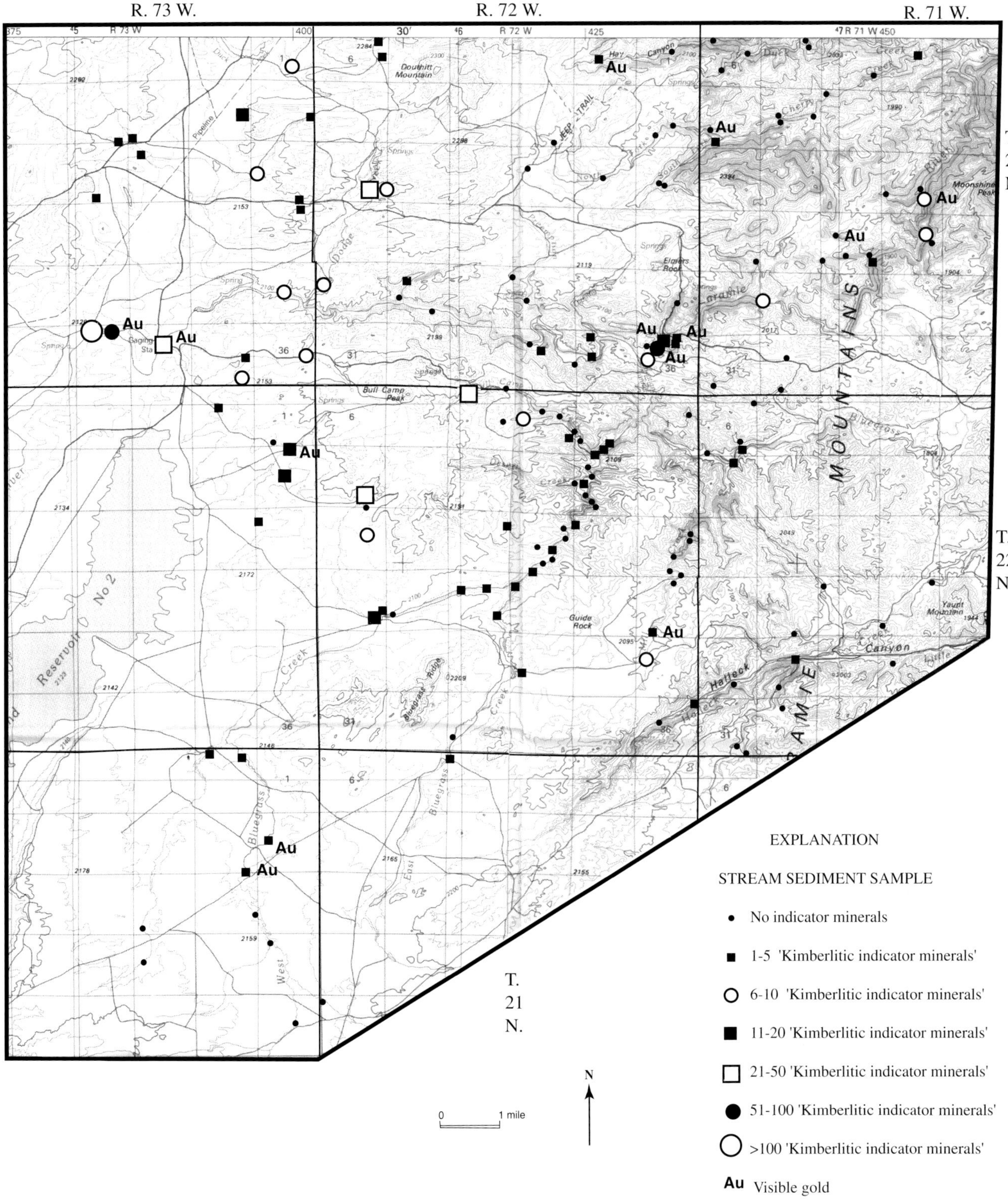

Figure 24. Map of the Elmers Rock greenstone belt showing locations of anomalous samples (after Hausel and others, 1988, 1997).

Seminoe Mountains region. In the Seminoe Mountains region north of the Medicine Bow Mountains, eight pyrope garnets and four chromian diopsides were recovered from a sample taken in a gold paleoplacer along the northeastern flank of the Seminoe Mountains greenstone belt (Hausel, 1993a). The source of the indicator minerals was not determined. In the same region, a circular depression

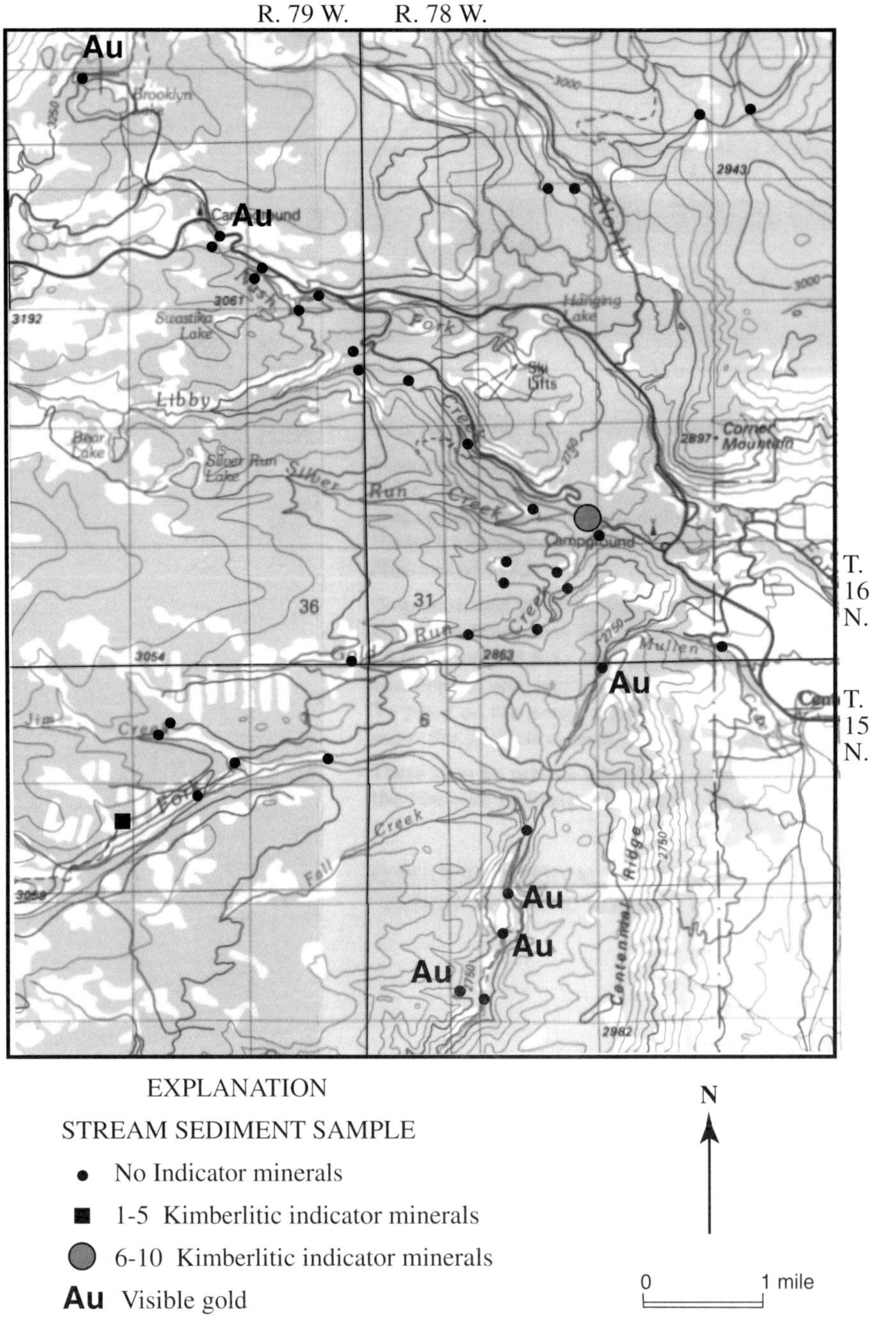

Figure 25. Map of a portion of the Medicine Bow Mountains showing locations of kimberlitic indicator mineral anomalies (after Hausel and others, 1988, 1997).

was identified by Woody Motten (personal communication, 1996) as a follow-up to the indicator minerals discovered by the Wyoming State Geological Survey. Geophysical surveys over the structure yielded a distinctive EM (electromagnetic) conductor and a weak magnetic anomaly. Modeling of the geophysical data supports the presence of a pipe-like structure at the site; however, no kimberlite or lamproite was found (Hausel, personal field notes, 1996).

Umbrella gravel pit. To the west of the Medicine Bow Mountains, a sample of alluvium collected by Bret Boundy (personal communication, 1995) from the Umbrella gravel pit near Dixon, along the western flank of the Sierra Madre, yielded a few small pyrope garnets.

Sage Creek Basin. In the Sage Creek Basin (sections 32 and 33, T17N, R88W) near the northern edge of the Sierra Madre, a fingernail-size diamond, estimated to have been more than 3 carats in weight, was found several years ago. The octahedron was verified by a gemologist and was found at the base of some aspen trees (Ralph Platt, personal communication, 1997).

Fremont County

A diamond was reportedly found in the late 1800s in the Beaver placers of the South Pass-Atlantic City mining district of the southern Wind River Range (Hausel, 1991). The diamond was reportedly sold for $1000.

Laramie County

Laramie County includes a portion of the eastern flank of the Laramie Mountains which includes the Iron Mountain kimberlite district (see Albany County, page 25).

Iron Mountain district. About 45 miles north of the State Line district, 57 kimberlite dikes and blows were mapped in the Iron Mountain district by Smith (1977). A few additional kimberlites were recently found by the author (personal field notes, 1997) and by Paul Graff and Gordon Marlatt (personal communication, 1998). This district is in the northwestern corner of Laramie County (**Figure 7**). Apparently only one kimberlite in this region was tested by Cominco American Incorporated, and no diamonds were found.

Peridotitic garnets recovered from the Iron Mountain kimberlites are chemically equivalent to G9 and G10 pyropes. The presence of G10 pyropes suggests that the kimberlites originated within the diamond stability field; however, ilmenite analyses from the kimberlites show depletion in magnesium and slight enrichment in chromium. Such depletion and enrichment are thought to correlate with the increased oxidation state of the kimberlite magma and imply that diamond preservation was unlikely (McCallum and Waldman, 1991). However, the ilmenite compositions from these kimberlites are similar to the ilmenite compositions in the kimberlites at Kelsey Lake where diamonds are well preserved (Coopersmith and Schulze, 1996). This suggests that the Iron Mountain district may deserve further investigation.

Stream sediment samples in the area identified kimberlitic indicator minerals: (1) associated with known kimberlites, and (2) in drainages unrelated to the known kimberlite occurrences (**Figure 26**).

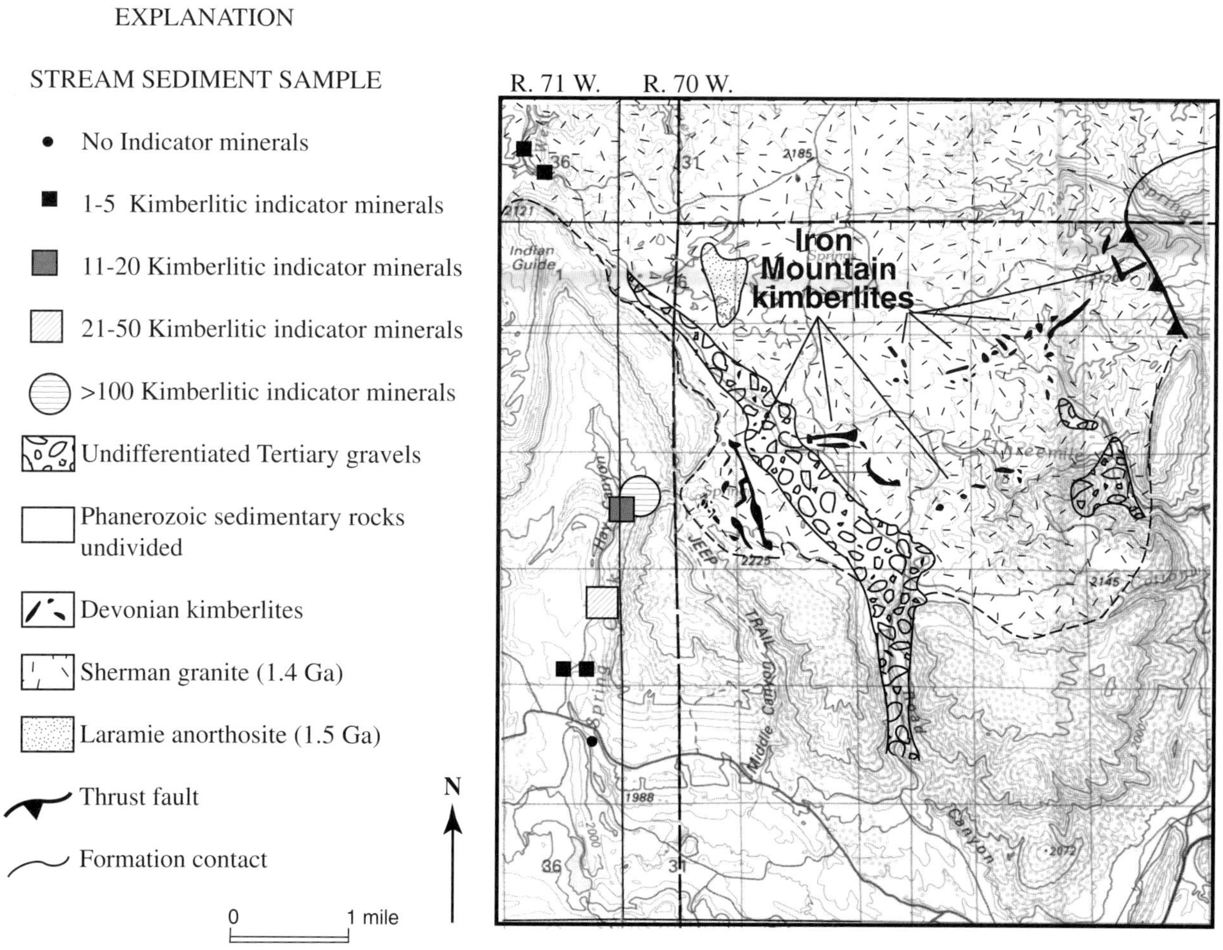

Figure 26. Generalized geologic map of the Iron Mountain kimberlite district showing locations of stream sediment samples (modified from Smith, 1977).

Two of the samples are located north of the known kimberlites and are separated from the intrusives by a topographic high. Other anomalies in the area include a group of topographic depressions west of the kimberlites in sections 8, 9, 16, and 17, T19N, R71W. EM and magnetic profiles completed by the Wyoming State Geological Survey over a few of the depressions were inconclusive.

Sublette County

A diamond about one inch in diameter was reportedly found by Snook Moore near Tourist Creek west of Gannett Peak in the Wind River Range. The diamond was later destroyed in a ranch fire (J. David Love, personal communication, 1981).

Sweetwater County

Sweetwater County in southwestern Wyoming contains one of the largest lamproite fields in the world as well as an extensive kimberlitic indicator mineral anomaly covering an area of 500 to 1000 mi^2 (**Figure 7**). The anomaly extends westward into adjoining Uinta County (see Uinta County, pg. 38) and southward into northern Utah and northwestern Colorado. This extensive anomaly suggests the presence of numerous undiscovered pipes.

The Green River Basin is a Laramide basin filled in part with lacustrine sedimentary rocks that crop out in the basin's center. The eastern edge of the basin is bordered by a large anticline known as the Rock Springs uplift (Late Cretaceous-Paleocene in age). The entire basin and Rock Springs uplift appear to be underlain by basement rocks of the Wyoming craton, and should be favorable for diamond exploration.

The lamproites of the Leucite Hills lie north of Rock Springs (**Figure 7**). This volcanic field includes several 3.1- to 1.4-Ma lamproites that lie along the northern flank of the Rock Springs uplift. The age dates show that the lamproites initially erupted in the extreme northwestern portion of the field and the eruptions progressed to the southeast through time.

The rocks that form the Leucite Hills consist of rare ultrapotassic mafic volcanic rocks known as wyomingite (diopside-leucite-phlogopite lamproite), madupite (diopside-madupitic lamproite), orendite (diopside-sanidine-phlogopite lamproite), and olivine orendite (olivine-sanidine-phlogopite lamproite). Geochemically, these lavas range from 42.65 to 56.34% SiO_2, 2.52 to 12.66% K_2O, and 5.8 to 12.75% MgO.

Individual flows vary in thickness from less than a foot to 122 feet and include vesicular lavas, scoria, autolithic intrusive breccias, lapilli tuffs, tuff breccias, and agglomerates. Although vents have been found in association with most flows, vent facies breccias are inconspicuous at South Table Mountain, North Table Mountain, Black Rock, and Hatcher Mesa. Either these vents were removed by erosion or they were buried by later flows (Ogden, 1979). Since it is not uncommon for lamproite flows to bury vent facies lamproites, the author suspects that the latter possibility is the more likely of the two.

Petrographic studies show the Leucite Hills lamproites to contain diopside, phlogopite, titanian-potassium-richterite, leucite, sanidine, wadeite, priderite, and/or olivine, with minor apatite, perovskite, ilmenite, armalcolite, and spinel. Kalsilite and primary analcite have also been identified in some lamproites. The typical kimberlitic indicator minerals (pyrope garnet, chromian diopside, and picroilmenite), which are rare in lamproites, have not been identified in the Leucite Hills lamproites. However, a sediment sample collected along the northern margin of Endlich Hill yielded one pyrope garnet (W.D. Hausel, personal field notes, 1984).

According to Carmichael (1967), the Leucite Hills volcanics exhibit similarities to the phlogopite lamproites and olivine lamproites in Western Australia and Murfreesburo, Arkansas. In particular, Carmichael (1967) noted similarities of the Leucite Hills madupites to Arkansas olivine lamproites as well as to kimberlite, and suggested that the Leucite Hills may have been derived by the fractionation of a kimberlitic liquid.

The noted similarities of lamproite and kimberlite should lead to some diamond exploration in the Leucite Hills; however, only minor activity has occurred. To date, one small sample of olivine-orendite lamproite breccia (<1 short ton) was collected from the northern edge of Endlich Hill and tested for diamonds by the Wyoming State Geological Survey (Hausel, Love, and Sutherland, 1995). Although no diamonds were recovered, the sample was too limited in size and extent to properly assess the diamond potential.

Another small grab sample collected from the Leucite Hills several years ago yielded one small, transparent octahedron. However, the octahedron was identified as spinel using x-ray precession (W.D. Hausel, personal notes, 1984).

Plate 3

Plate 3a. Open pit operations at the Kelsey Lake diamond mine.

Plate 3b. View of the diatreme facies kimberlite in a highwall in one of the pits at Kelsey Lake showing where the kimberlite stoped under the granite.

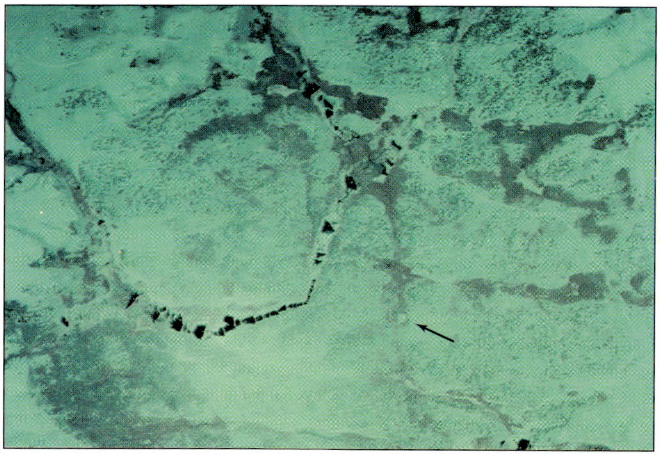

Plate 3c. False-color infrared aerial photograph over the Eagle Rock area, Laramie Mountains. On the left side of the photograph, a prominent, circular drainage is visible with several small beaver ponds (black). Several kimberlitic indicator mineral anomalies were identified in association with these ponds (refer to Figure 20). Near the center of the photograph, immediately right of the beaver ponds, a north-northwest-trending stand of aspen trees (red) encloses a small, circular, open area known as the Eagle Rock anomaly (indicated by arrow). The source of this vegetation anomaly remains unknown.

Plate 3d. Kimberlitic indicator minerals found on an anthill near Butcherknife Draw in the Green River Basin.

Plate 3e. Kimberlitic indicator minerals recovered from the DK breccia pipe, Cedar Mountain (photograph by Richard Kucera).

Some limited exploration has also been conducted in the Leucite Hills by a few companies in recent years.

Of the known lamproites in the world, olivine lamproites offer the highest potential for significant diamond mineralization. Thus, exploration should focus on the testing of the olivine orendites in the Leucite Hills, and on a search for buried lamproite. Olivine-rich lamproites tend to erode rapidly due to serpentinization of olivine and typically merge into the surrounding topography.

Airborne geophysical and remote sensing surveys may provide a means to search for buried lamproite. Both types of surveys were used successfully in the Colorado-Wyoming State Line district in the search for kimberlite, and one might expect similar successes in the Leucite Hills, as well as in the remainder of the Green River Basin.

Since the madupitic lamproites contain magnetite, these lavas should yield detectable magnetic anomalies, while other lamproites may only yield weak magnetic anomalies. The olivine lamproite at Prairie Creek, Arkansas, for example, is magnetic and produced a distinct dipolar magnetic anomaly (Bolivar and Brookins, 1979). However, rock samples collected by the author from the Argyle olivine lamproite in Western Australia are not magnetic to very weakly magnetic. The tuff phases of the pipe at Argyle are weakly magnetic (Atkinson, 1989).

West of Argyle, airborne magnetic and radiometric surveys were used successfully in the discovery of diamondiferous lamproite in the Ellendale field, west Kimberley (Jaques and others, 1986). In addition, INPUT® surveys, which were used successfully to detect several diamondiferous kimberlites in the Colorado-Wyoming State Line district (Paterson and MacFayden, 1984), have also been used successfully in the Ellendale lamproite field (Atkinson, 1989).

Remote sensing surveys may also provide useful data, since lamproites and kimberlites are emplaced along fractures, and many of these structures should be detectable on aerial photography. For instance, in the Colorado-Wyoming State Line district, Hausel and others (1979) were able to identify the controlling structures for kimberlites through geologic mapping and air photo reconnaissance. Air photographs were also used to identify the controlling structure for the Cedar Mountain breccia pipes in the southern Green River Basin (Richard Kucera, personal communication, 1996). In addition, several of the kimberlite intrusives in the State Line district were isolated using various color-band composites from the unique reflectance of the weathered kimberlite (Marrs and others, 1984; Marks, 1985). It is not known if such surveys could be used to locate lamproites in the Leucite Hills, but because of the unique chemistry of these rocks, it is likely that they would be detectable on various composites.

Leucite Hills

In total, the Leucite Hills area contains 22 individual volcanic features, including flows, dikes, necks, plugs, and cinder and pumice cones, as mapped by Ogden (1979). These known lamproites in the Leucite Hills (**Figure 27**) are described below.

Black Rock. Black Rock is the easternmost lamproite of the Leucite Hills (**Figure 28**). Cross (1897) suggested that Black Rock represented a plug; however, Smithson (1959) and Ogden (1979) indicated Black Rock was a basal pyroclastic olivine orendite overlain by a lava flow, and that there was no evidence of a plug. Possibly, the lava flow obscures hidden vent facies volcaniclastics. The top 80 feet of the mesa consists of alternating layers of vesicular and non-vesicular lava.

Dunite nodules, and two generations of olivine phenocrysts, have been identified at Black Rock (Speer, 1985). One phenocryst is surrounded by a reddish reaction rim, has 0.3% CaO, and a composition of Fo_{90} (90% forsterite). The other phenocryst has 0.2% CaO and a composition of Fo_{86} (Carmichael, 1967). Recently, the author collected more than 10,000 carats of olivine, much of it transparent, gem quality peridot from anthills and rocks along the margin of Black Rock. The grains from the anthills ranged from 2 to 8mm. However, olivine megacrysts more than 1 inch across were recovered from the Black Rock lamproite.

Boars Tusk. Boars Tusk forms a prominent volcanic neck composed of wyomingite (phlogopite-leucite lamproite) that rises more than 300 feet above the valley floor. The agglomerate contains common shale and sandstone xenoliths from the Green River and Wasatch Formations (Tertiary), with occasional granite xenoliths and autobrecciated fragments of lamproite.

Another remnant of a volcanic neck, known as Matthews Hill, forms a small, low-lying hill a short distance south of Boars Tusk (Ogden, 1979).

Cabin Butte. Cabin Butte consists of an orendite lava flow with a volcanic plug dome along its northern end (Ogden, 1979).

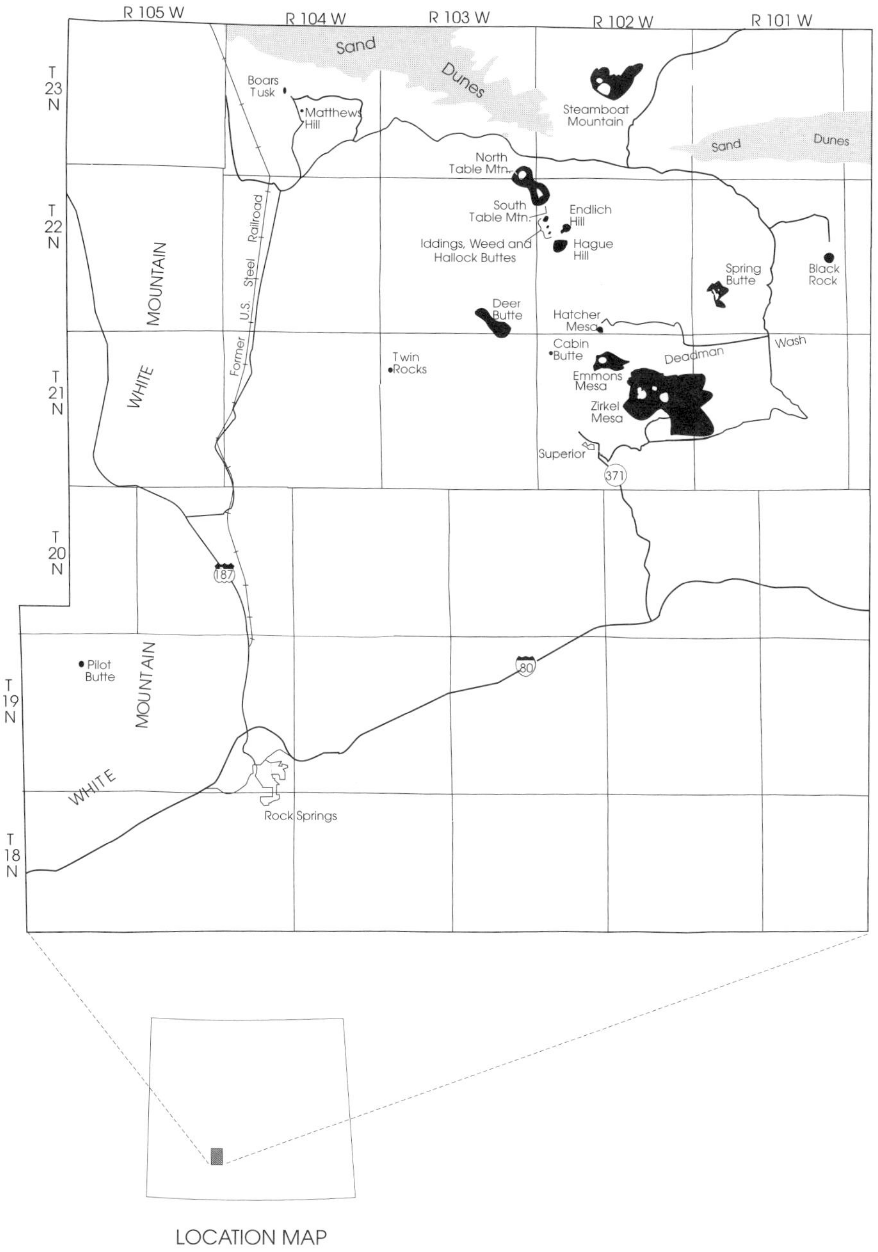

Figure 27. Location map of the Leucite Hills lamproite field, southwestern Wyoming.

Deer Butte. Deer Butte is a remnant of a composite volcanic center which includes two cinder cones, three separate lava flows, and a vent exposed in its southern wall. The scoria of Deer Butte is composed of wyomingite, while the lava is predominately orendite (Ogden, 1979).

Emmons Mesa. Emmons Mesa, formed predominantly of orendite, includes a volcanic cone and at least two flows, including an overlapping flow that predates the cone. The cone contains xenoliths of a rock formed of pyroxene, plagioclase, and phlogopite, as well as fragments of amphibolite, anorthosite, granulite, granite, and sedimentary rocks. The flow contains granite and sedimentary xenoliths (Ogden, 1979).

Endlich Hill. Endlich Hill includes both flows and volcaniclastics composed of olivine orendite. More than one population of olivine has been recognized at Endlich Hill (Cross, 1897; Carmichael, 1967). In addition to anhedral xenocrysts and subhedral to euhedral microphenocrysts, the lamproites enclose anhedral olivine mantled by phlogopite laths that are in disequilibrium with the host magma. These anhedral olivines possibly represent upper mantle-derived xenocrysts (Carmichael, 1967).

Hague Hill. Hague Hill forms a low-lying hill south of Endlich Hill. The lavas at Hague Hill are composed of orendite (Ogden, 1979).

Hatcher Mesa. According to Ogden (1979), Hatcher Mesa is an eroded remnant of a small ponded lava flow. The center of the mesa is depressed by 6 feet relative to the edges, indicating the surface sagged as the lava flowed back into the vent. The vent is not conspicuous. According to Smithson (1959), Hatcher Mesa is composed primarily of orendite with wyomingite present in the basal part of the flows.

The mesa is composed of light- to brownish-gray vesicular lava with small 2-mm phlogopite grains and black to dark-green 2.5-mm clinopyroxene grains. In thin section, microphenocrysts of clinopyroxene, phlogopite, apatite, and iron-titanium oxide with uncommon olivine occur in a groundmass of leucite, clinopyroxene, potassium richterite, apatite, priderite, wadeite, and glass (the absence of sanidine in these rocks distinguishes them as wyomingite). The olivine grains are unzoned and characterized by high $Mg/(Mg+\Sigma Fe^{2+})$ and Ni. In these respects, the olivines are comparable to olivines from lherzolite and harzburgite mantle nodules in kimberlite (Barton and van Bergen, 1981).

Whole rock analyses show relatively high K_2O, BaO, SrO, P_2O_5, and ZrO, and relatively low Na_2O

Figure 28. Black Rock lamproite mesa, view to northeast from Spring Butte.

and Al_2O_3 compared to a typical mafic volcanic rock (Barton and van Bergen, 1981). The lavas contain the most abundant and diverse assemblage of xenoliths and xenocrysts in the Leucite Hills. Xenoliths from this locality include rocks of both mafic and ultramafic affinities. According to Smithson (1959) the xenolith assemblage includes granite, norite, and mica schist. Inclusions of phlogopite, apatite, and glass similar to those at Black Rock are found at Hatcher Mesa.

Iddings, Weed, and Hallock Buttes. These are three small knobs of wyomingite lava that lie along a N40°W trend between Hague Hill and South Table Mountain. The buttes are apophyses originating from a dike (the Wortman dike) in the subsurface. Iddings Butte is zoned and has an interior of dense lava with an exterior of volcanic agglomerate (Ogden, 1979).

Pilot Butte. Pilot Butte is the southwesternmost lamproite in the Leucite Hills. The madupitic lavas from Pilot Butte are silica-poor and magnesian-rich relative to other lavas in the Leucite Hills. These rocks are enriched in Ba, Sr, Th, and REE (rare earth elements); are depleted in Ni; and have lower K/Rb ratios than most phlogopitic lamproites, all of which suggest the madupitic lamproites at Pilot Butte may be derivatives from other phlogopitic lamproites (Mitchell and Bergman, 1991). Spinels from the lavas at Pilot Butte are chrome-poor (>1 wt.% Cr_2O_3), magnesian (<5 wt.% MgO), titaniferous (>10 wt.% TiO_2) magnetites, and are unlike the chrome-spinels (9 to 58 wt.% Cr_2O_3) found in the diamondiferous lamproites in Western Australia and Murfreesburo, Arkansas. The lack of chrome spinels at Pilot Butte suggest that this lamproite is a poor target for diamond.

Spring Butte. Spring Butte (originally named Orenda Mesa) is considered the type locality for orendite. The butte is a compound volcanic center with six cinder cones, at least six flows, and three dikes. Flow thicknesses range from 10 to 77 feet. The cones consist of welded clastic flows with ribbon and breadcrust bombs. Smithson (1959) reported that Fort Union Formation sandstones with some rare anorthosites occurred as xenoliths at Spring Butte.

Steamboat Mountain. Rocks from Steamboat Mountain are predominantly orendite aphanitic vesicular flows containing some sedimentary xenoliths. According to Carmichael (1967), a glassy wyomingite sample collected from this mesa yielded 12.66% K_2O, indicating this is one of the most potassic lavas in the world.

Table Mountain. Table Mountain (**Figure 29**) consists of North Table Mountain, South Table Mountain, and Middle Table Mountain. North Table Mountain is a volcanic mesa consisting of a single flow with well-developed flow layering of light gray orendite and dark gray wyomingite. South Table Mountain forms a prominent mesa next to North Table Mountain, and consists of olivine orendite with some wyomingite. On the southern margin of North Table Mountain, a small volcanic plug known

Figure 29. North and South Table Mountain lamproites, on the left skyline, and Steamboat Mountain on the right skyline. View is to the northwest from Spring Butte.

as Middle Table Mountain occurs with greenish layers of madupite alternating with brownish layers of wyomingite (Ogden, 1979).

The flows on South Table Mountain are relatively MgO-rich due to the presence of phenocrystal and xenocrystal olivine. Small dunite xenoliths have also been reported in these rocks. The madupitic lavas from Middle Table Mountain are silica-poor and magnesian-rich relative to other lavas in the Leucite Hills and contain shale xenoliths (Speer, 1985).

Twin Rocks. Twin Rocks (locally known as the Badgers Teeth) form five small projections of xenolith-rich lava that rise from an inverted cone along an east-west trend. These are part of a volcanic neck conduit which rose as apophyses from a subsurface dike. The bulk of the neck is formed from autobrecciated agglomerate. Texturally these rocks are wyomingites, but chemically they are similar to madupite (Ogden, 1979).

Zirkel Mesa. Zirkel Mesa is the largest lamproite exposure in the Leucite Hills and is composed primarily of orendite. Flow thickness varies from less than 3 feet to as much as 50 feet (Kuehner, 1980). Highly vesicular, pumiceous leucite lamproite exposed in a quarry along the eastern edge of Zirkel Mesa exhibits flow banding with common country rock xenoliths (Hausel, Sutherland, and R.W. Gregory, 1995).

Ogden (1979) reported that the mesa consisted of six cinder cones rising more than 240 feet above the mesa surface. These cones coalesce into a single, volcanic-capped plateau.

Green River Basin anomaly

Southwest of the Leucite Hills, pyrope garnet, chromian diopside, and other mantle-derived kimberlitic indicator minerals, are found in numerous anthills and adjacent soils north of Cedar Mountain (**Plate 3d**) (McCandless, 1982, 1984); similar indicator minerals have also been found in the basal conglomerate of the Bishop Conglomerate (Oligocene) on Cedar Mountain (**Figure 30**), Sage Creek Mountain, and in Colorado on Diamond Peak (McCandless and others, 1995) (**Figure 31**). This anomaly is widespread, covering 500 to 1000 mi^2 in

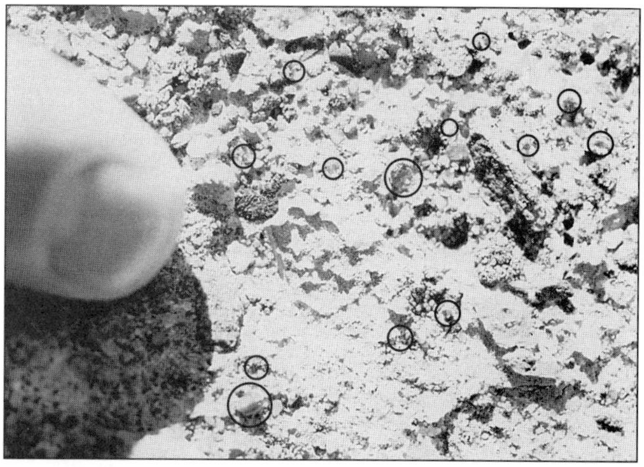

Figure 30. Small chromian diopside crystals (circled darker areas that are green on a color photograph) in whitish-colored matrix of Bishop Conglomerate from Cedar Mountain.

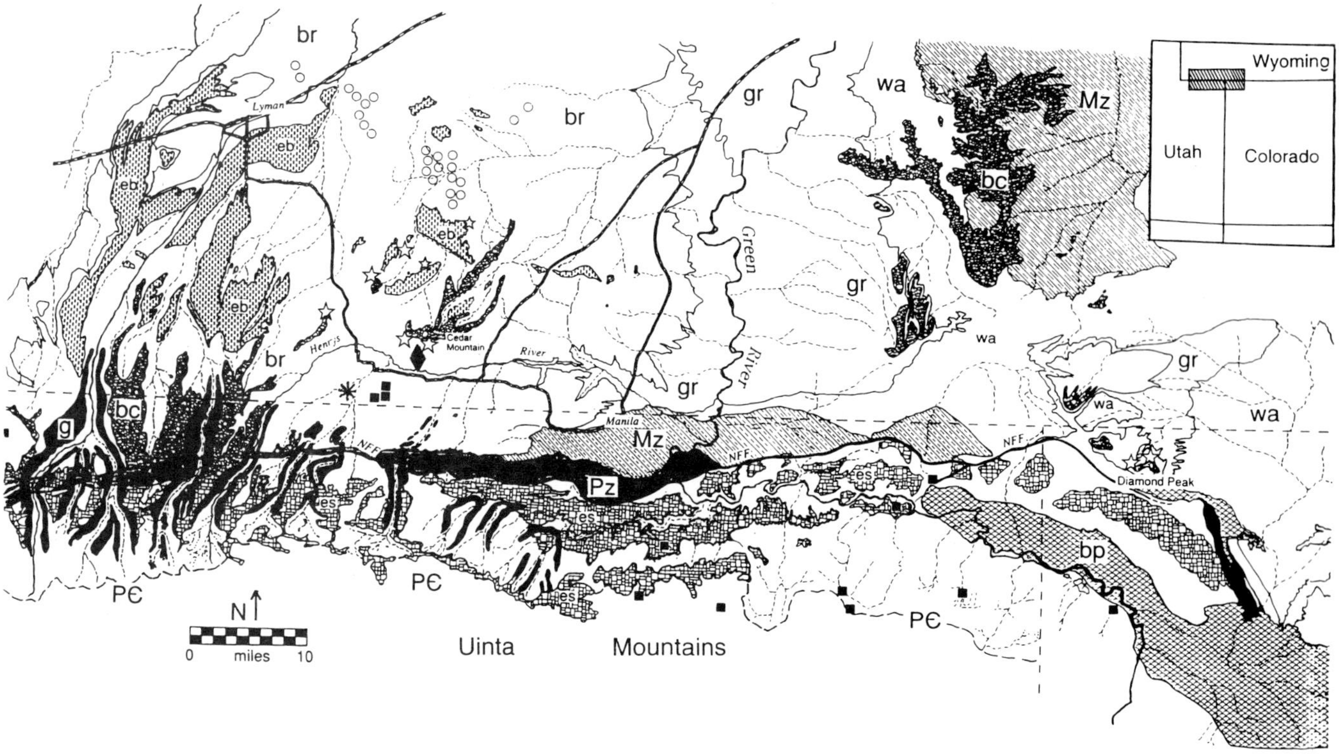

Figure 31. Simplified geologic map of the southern Green River Basin and north slope of the Uinta Mountains. Pre-Tertiary units are Uinta Mountain group (PЄ), Paleozoic (Pz) and Mesozoic (Mz) sedimentary rocks. Tertiary sedimentary units are Green River Formation (gr), Bridger Formation (br), and Wasatch Formation (wa). Erosion surfaces are (es) earliest unit capping the Uinta Mountains, followed by Bishop Conglomerate (bc), erosional surfaces later than the Bishop Conglomerate (eb), and the Browns Park Formation (bp). Pleistocene glacial moraines are (g). The projected trace of the North Flank Fault (NFF) is shown as solid line. Flaming Gorge reservoir is removed for clarity. Kimberlitic indicator minerals are found in ant mounds (circles), pediments and conglomerates (stars), and streams (squares). Approximate locations of the Cedar Mountain breccia pipes (diamond symbol) and Lone Tree anomalies (asterisk) are shown. Modified from McCandless and others (1995).

the Green River Basin of Sweetwater and Uinta Counties and extending into northwestern Colorado and extreme northeastern Utah. The chemistry of the indicator minerals is not favorable for diamond and suggests that the indicator minerals were derived from a source in the upper mantle within the graphite stability field (McCandless and others, 1995).

Within the anomalous area, an oil well drilled in section 19, T15N, R111W, reportedly cut a trona-bearing unit at a depth of 2700 feet. According to the geologist's log, a bed in this unit contained talc, chlorite, saponite, green hornblende, and pyroxene (Love, 1963). These minerals are suggestive of a mafic to ultramafic parental rock.

Four diamonds were reportedly found in Sweetwater County in the Butcherknife Draw area by George Stevens (Paul and Jean Miller, personal communication, 1994). The Butcherknife Draw area lies within the anomalous area. One of the diamonds, currently owned by a Green River resident, was cut in Germany and mounted in a ring. The stone measures 0.3 inch across and was apparently confirmed as diamond by a gemologist with a GEM® diamond tester (Wayne Sutherland, personal communication, 1995).

Teton County

A 7- to 9-carat blue-white gem-quality diamond was reportedly found in a prospect pit in the Gros Ventre Range (J. David Love, personal communication, 1981).

Uinta County (see also Sweetwater County)

The Green River Basin kimberlitic indicator mineral anomaly extends westward from Sweetwater County into Uinta County (**Figure 7**). Mantle-derived minerals are found in the Bishop Conglomerate along the western flank of Cedar Mountain (sections 16, 17, 18, 19, and 20, T13N, R112W) and on Sage Creek Mountain to the northwest (sections 34 and 35, T14N, R113W, and section 3, T13N, R113W). Similar minerals are also reported in soils on the surface and in road cuts in sections 11, 12, 13, 14, 15, 21, 22 and 23, T13N,

R113W, south of Sage Creek Mountain (Gordon Marlatt, personal communication, 1994). This area also includes a group of circular anomalies of unknown origin west of Lonetree and a group of breccia pipes and dikes along the flank of Cedar Mountain.

Lonetree anomalies. A group of circular topographic features located in sections 7, T12N, R113W, and sections 12, 13, and 24, T12N, R114W, were identified southwest of the Green River Basin kimberlitic indicator mineral anomaly near Lonetree (Gordon Marlatt, personal communication, 1994). Ground EM and magnetic surveys over some of these features detected conductors and very weak magnetic anomalies of unknown origin (Hausel and Gregory, 1996).

Cedar Mountain breccia pipes. East of Lonetree, five diamonds were reportedly discovered in the early 1980s in a drainage on the flank of Cedar Mountain. Along this drainage 10 mafic to ultramafic breccia pipes and dikes were discovered in 1995 (**Figure 32**) (Richard Kucera, personal communication, 1995). The pipes lie along a 5- to 10-mile-long, northerly trending lineament in the Bridger Formation (Eocene). Geophysical surveys conducted over the pipe showed the pipe breccia to be conductive, although the conductivity of the surrounding clay-rich country rock of the Bridger Formation was higher than the pipe, resulting in a relative conductivity low over the pipe. Magnetic surveys yielded only weak, positive anomalies associated with the DK pipe. Recently, Guardian Resources reported the discovery of two more breccia pipes in the basin (Press Release, Guardian Resources, 1997).

Cratonic rocks are completely buried in this area, and are only exposed as xenoliths and xenocrysts in the breccia pipes. Other mantle xenoliths and xenocrysts occur in the lamproitic breccias and flows in the Leucite Hills 60 miles northeast of Cedar Mountain.

The Cedar Mountain pipes and dikes host kimberlitic indicator minerals that are geochemically similar to those found in the Bishop Conglomerate and in anthills to the north (**Plate 3e**). The presence of these minerals in the pipes provides a source for some of the detrital minerals found; however, the pipes can only account for a small portion of the indicator minerals in this region. Thus, to explain the widespread anomaly one must consider the possibility of numerous undiscovered pipes in the rest of the basin and in the Uinta Mountains to the south.

The largest diatreme in the DK group near Cedar Mountain measures 400 by 700 feet. The breccias forming these pipes are greenish-gray and contain abundant lithic sedimentary fragments from the Bridger Formation that were trapped in the pipes from the underlying strata. The breccias also include crystalline basement fragments and mantle-derived eclogite nodules.

Kimberlitic indicator minerals are abundant in the pipes, and the size of these grains is equivalent to the size of the detrital mineral grains found throughout the basin, further suggesting the presence of undiscovered pipes. The kimberlitic indicator minerals recovered from the DK pipes and dikes include garnets, pyroxenes, and oxides, along with some mini-xenoliths. Both peridotitic and eclogitic garnets occur in the mineral assemblage. The peridotitic garnets are dominantly calcic pyropes (G9), and the eclogitic pyrope-almandine garnets are low in sodium (≤0.06 wt.% Na_2O). All but one pyroxene yielded K_2O < 0.07%. The exception was a calcic chromian diopside of peridotitic paragenesis that yielded 0.10% K_2O, suggesting a probable origin within the diamond cogenetic field (T.E. McCandless, written communication, 1996).

Oxides include both ilmenites and chromites. The ilmenites yielded 4.8 to 7.0 wt.% MgO and 0.0 to 4.5 wt.% Cr_2O_3, typical of kimberlitic ilmenites (the ilmenite compositions suggest favorable conditions for diamond preservation). The chrome spinels yielded 14.5 to 57.0 wt.% Cr_2O_3 and 0.0 to 19.7 wt.% MgO.

Seventeen garnet-pyroxene mini-xenoliths that were collected from the DK pipe were analyzed. Thirteen contained garnets with high almandine content suggesting a granulite (lower crustal) source. Four of the xenoliths yielded magnesian pyrope-almandine garnets of probable eclogitic origin. None of the garnets contained elevated Na_2O, which suggests that they originated above the diamond stability field (T.E. McCandless, written communication, 1996).

The host rock of the breccia pipe is unusual; because of the abundance of crustal xenoliths, good samples were difficult to obtain for whole rock analysis. Selected samples with minor xenolith content yielded (on a weight percent basis) 44.08 to 67.66% SiO_2, 0.27 to 0.47% TiO_2, 5.25 to 9.86% Al_2O_3, 2.82 to 5.13% Fe_2O_3, 0.04 to 0.22% MnO, 4.29 to 14.95% MgO, 5.10 to 17.99% CaO, 0.57 to 1.75% Na_2O, 1.41 to 2.49% K_2O, 0.22 to 0.56% P_2O_5, 6.6 to 17.77% LOI (Loss on ignition), 0.07 to 0.14% Cr_2O_3, and 256 to

a.

c.

b.

Figure 32. (a) Ultramafic to mafic dike emplaced in rocks of the Bridger Formation along the southwestern flank of Cedar Mountain; (b) mantle-derived breccia pipe exposed in dozer cut along Cedar Mountain (3-foot sledge hammer handle for scale); and (c) hand sample of the breccia at Cedar Mountain.

755 ppm Ni; one sample yielded 5.44% CO_2 (Hausel and others, 1996). The MgO content is low, the SiO_2 content is high compared to an average kimberlite, and the chemistry does not appear to be comparable to lamproite. However, the high nickel and chrome values suggest contribution from a mafic to ultramafic source rock.

Thin sections show the rock matrix to consist of cryptocrystalline silica. Back-scatter microprobe analyses of rock samples support the petrographic work and show the groundmass to be highly silicified (Tony Irving, personal communication, 1996). Silicification is also indicated by whole rock analyses which yielded high silica contents. These rocks contain no identifiable magmatic components in the matrix and may be suggestive of a gas-solid mixture that disrupted the overlying Bridger Formation.

Guardian Enterprises completed a six-hole drilling program on the DK pipe in early 1996. A total of 1250 feet of core was recovered, and a 3000-pound bulk sample was collected. The bulk sample was processed by the Wyoming State Geological Survey and the concentrates were run across a grease table; no diamonds were recovered (**Figure 33**).

Another sample collected by Guardian Resources was processed by the Saskatchewan Research Council. This 200-pound core sample reportedly yielded two diamonds, and a third diamond was apparently recovered from concentrates of a sample collected in alluvial material adjacent to one of the pipes. The alluvial diamond was reported to have a maximum diameter of 2.2 mm with SI clarity and J color (see Bruton, 1979, p. 275 and p. 289 for these grade systems). The diamonds from the core were reported to have maximum diameters of 1.07 mm and 0.13 mm, respectively (Guardian Resources, Ltd., Press Release, February 28, 1996).

Figure 33. Grease table in operation at the Wyoming State Geological Survey in Laramie. Since diamonds are hydrophobic (repel water), they are attracted to grease. The grease used in this process to attract the diamonds consists of 1 part paraffin and 10 parts Vaseline®.

According to a later press release from Guardian Resources (September 24, 1996), several more alluvial diamonds were apparently recovered from a new source other than the breccia pipes. Forty-eight diamonds were reported from the new alluvial source from an undisclosed location. The diamonds reportedly weighed a total of 2.3 carats. However, it should be noted that none of the diamonds have been confirmed by the Wyoming State Geological Survey.

Bighorn Basin

Several kimberlitic indicator minerals, including some pyropes with G10 compositions, were reportedly identified in conglomeratic sources in areas of Precambrian and Paleozoic exposures along the Bighorn and Owl Creek Mountains of northern Wyoming by a diamond exploration company several years ago (Chuck Mabarak, personal communication, 1997). Apparently, the G10 garnets were found in the most westerly exposures of the conglomerates (pyropes designated as G10 have compositions similar to diamond-inclusion garnets).

Montana (Wyoming craton)

Alnöites, monticellite peridotites, carbonate-rich mica peridotites, monchiquites, lamproites, and kimberlites are all reported in Montana. Many of these intrusives occur in the central Montana alkalic province in eastern Montana, which overlies the Wyoming Province (**Figure 34**) (Hearn, 1989; Scambos, 1991). A few lamproites have also been found in western Montana (Pete Ellsworth, personal communication, 1996) and lie off the craton.

In eastern Montana, mantle-derived breccias and dikes are described in the Missouri Breaks, Porcupine Dome, and Grassrange areas. More than 30 carbonate-rich dikes and breccia pipes are reported in the Porcupine Dome and Grassrange areas, and a group of kimberlites occur in the Missouri Breaks area.

A few diamonds have been reported in western Montana including some relatively large stones of unknown origin. Other diamonds have been reported from the Sweet Grass Hills in northern Montana, just south of the Alberta border. With the

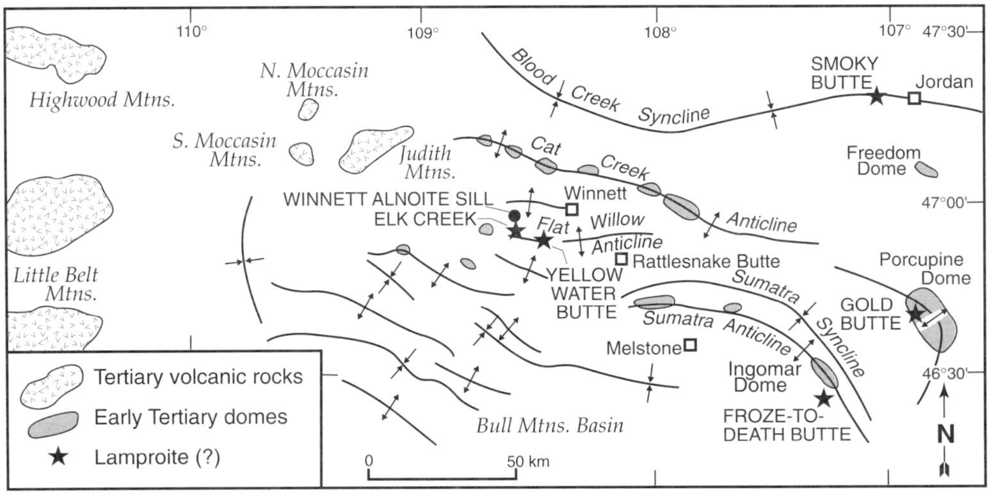

Figure 34. Regional map of Tertiary structures, lamproites, and lamprophyre occurrences near Jordan, Montana (after Mitchell and Bergman, 1991).

presence of a cratonic basement underlying much of Montana, as well as favorable host rocks, Montana would be considered high priority for diamond exploration.

Beaverhead County

A diamond was reported from Grasshopper Creek southwest of Dillon in the southwestern corner of the state (Sinkankas, 1959). No details were given.

Blaine and Hill Counties

Two diamonds were reportedly found in gravels of the Etzikom Coulee in the Milk River drainage north of the Sweet Grass Hills. The diamonds weighed 0.14 and 0.17 carats, respectively (Lopez, 1995). The placer diamond occurrence lies near a buried magnetic anomaly aligned with presumed lamproitic rocks in Alberta, Canada, where considerable diamond exploration activity has occurred in recent years (D.A. Lopez, personal communication, 1997).

Garfield County

A lamproite complex, known as the Smoky Butte lamproite (27 Ma), lies 6.5 miles west of Jordan in the Missouri Breaks region of northeastern Montana. The complex consists of several lamproite exposures along a N30°E trend (**Figure 35**). The lamproites are exposed at Radial Dike Butte, Bull Snake Knob, Half Sediment Butte, Smoky Butte, Instrument Butte, Ship Rock, and Wall Rock (**Figure 36**).

The lamproites are sanidine-diopside-richterite-phlogopite lamproites. These lie along a dike that swells to 122 feet in width. The rocks include vesicular, massive, glassy, hypabyssal breccias, with minor tuffs and pyroclastics. Smoky Butte is a vent (Mitchell and Bergman, 1991). Fipke and others (1995) reported that a few chromites from the Smoky Butte intrusive yielded geochemistries similar to diamond inclusion chromites. A small (65 kg) sample, however, yielded no diamonds.

Figure 35. A portion of the dike complex at Smoky Butte.

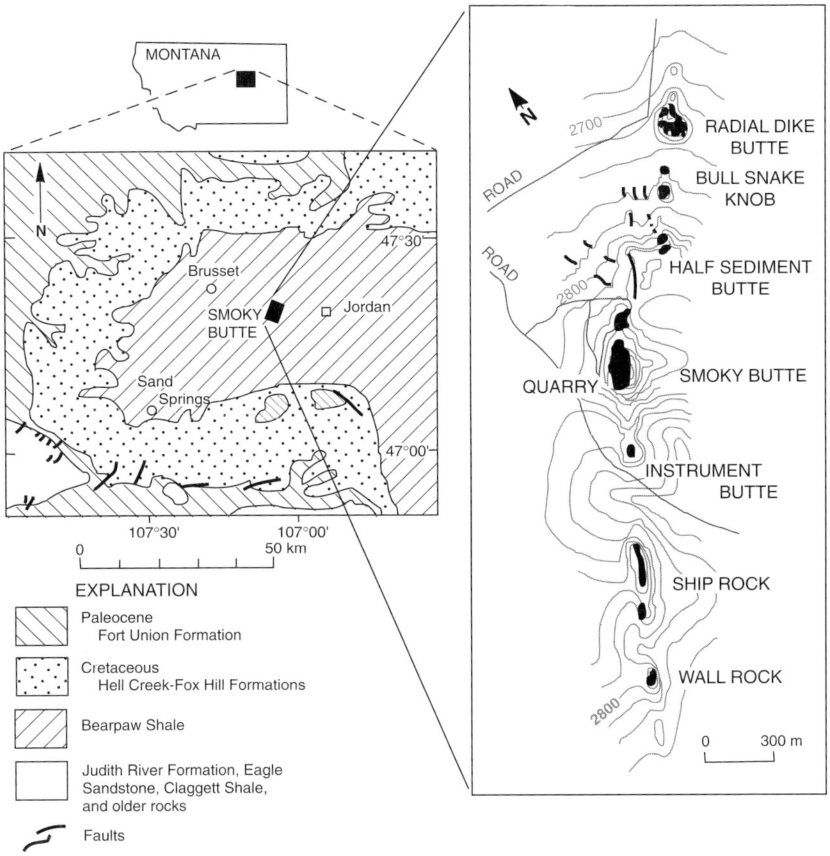

Figure 36. Map showing lamproite occurrences in the Smoky Butte lamproite field (after Mitchell and Bergman, 1991).

Glacier County

In 1883, a 12-grain diamond was discovered in the placer workings near Nelson Hill in the town of Blackfoot, northwestern Montana. The stone was described as a colorless dodecahedron with triangular markings (Kunz, 1885). In 1894, a 0.22-carat flawed stone was also reported from the county (Sinkankas, 1959).

Granite County

A phlogopitic lamproite flow, known as the Ruby Slipper (Bearmouth pipe), intrudes Devonian age Jefferson Formation and Tertiary basalts and rhyolites near Bearmouth along Bear Gulch in the Garnet Range. The flow is found in sections 3, 4, and 10, T11N, R14W.

The exposed plug consists entirely of flows which buried the underlying volcaniclastics. The underlying breccias were intersected by drilling (Pete Ellsworth, personal communication, 1996). The flow was initially mapped as a 1800-foot-wide leucite basalt by Carter (1982), but later identified as lamproite. According to Ellsworth (1996), the pipe forms a prominent 1500-foot circular knob, with crater facies volcaniclastics covered by talus and colluvium from the lamproite porphyry. Four other lamproitic prospects have been located in the region (Pete Ellsworth, personal communication, 1996).

Lewis and Clark County

In 1990, a jogger reportedly found an uncut, 14-carat diamond southwest of Great Falls near the town of Craig in west-central Montana. The stone, named the Lewis and Clark diamond, was sold for $80,000 (Anonymous, 1990a). Other diamonds have been found along the Missouri River to the south near Helena, including another relatively large diamond reported to weigh 8 carats (Anonymous, 1990b).

The only known intrusive with upper mantle material within this area lies along the eastern shoreline of Holter Lake on the Missouri River northeast of Helena. The intrusive, known as the Ming Bar monchiquite, contains abundant cognate metacrysts of olivine, pyroxene, and chromite with granulite, dunite, pyroxenite, and peridotite nodules (Meen and others, 1986). The intrusive originated in the upper mantle, but probably at too shallow a depth for diamonds.

However, some diamonds were apparently found in the vicinity of Ming Bar. Five diamonds were reportedly recovered from placer operations at Spokane Bar, and a 5-carat diamond was reported from Metropolitan Bar. Both placer localities lie upstream from Ming Bar (Tony Irving, personal communication, 1994).

Madison County

Diamonds have been reported from Greenhorn Gulch in the Greenhorn Range east of Dillon, southwestern Montana (Sinkankas, 1959). No details were given.

Petroleum and Fergus Counties (Grassrange Field)

A belt of ultramafic lamprophyres form the Grassrange Field approximately 4 to 6 miles south of Winnet, near Yellow Water Reservoir in east-central Montana. According to Doden (1996, 1997), the belt consists of ultramafic lamprophyric diatremes and dikes that were originally described as lamproites by Mitchell and Bergman (1991) (**Figure 37**).

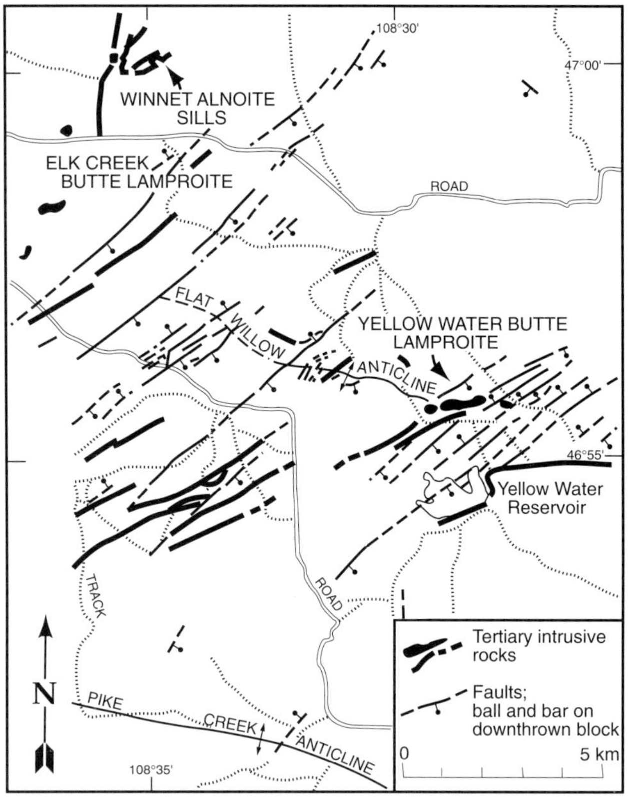

Figure 37. Location and structural map of the Yellow Water Butte and Elk Creek Butte lamproites (ultramafic lamprophyres?) (after Mitchell and Bergman, 1991).

Elk Creek Butte and Yellow Water Butte, located in the center of the field, form large ultramafic lamprophyres rather than true lamproites. The Yellow Water Butte diatreme, in particular, is very carbonate rich. These two buttes lie within a group of ultramafic intrusives known as the Winnet sills, which are nepheline-haüyne alnöites, a carbonate-free rock (Doden, 1996).

The breccias that form Yellow Water Butte consist of massive to brecciated, olivine-phlogopite-diopside-carbonate lamprophyre (?) and massive hypabyssal olivine lamprophyre (?). The Elk Creek vent-dike complex consists of intrusive breccia and lapilli tuffs (Mitchell and Bergman, 1991). Similar, but less extensive outcrops also occur in the area (Hausel, personal field notes, 1994). Doden (1996) suggested that the Yellow Water Butte breccias were part of a highly gas-charged magma that rapidly ascended from the mantle and was virtually unaffected by crustal contamination.

Phillips County

In Phillips County north of the Grassrange Field, a group of intrusives, known as the Williams kimberlites, occurs near Landusky. The Williams kimberlites form a group of four closely spaced diatremes (**Figure 38**) in the eastern part of an east-northeasterly trending swarm of ultramafic alkalic diatremes, dikes, and plugs (46 to 51 Ma) in the Missouri Breaks area of north-central Montana. These rocks preceded the intrusion of Smoky Butte (27 Ma) located 67 miles to the southeast. The rocks contain a host of xenoliths including garnet-bearing lherzolites, harzburgites, and dunites. The Williams 1 diatreme occupies a surface area of about 750 by 105 feet, and contains the typical kimberlitic indicator minerals pyrope garnet, chromian diopside, and magnesian ilmenite.

The Williams 2 kimberlite has a surface area covering about 128 by 375 feet, and has a zone of kimberlite breccia with abundant Paleozoic limestone and dolomite xenoliths. The Williams 3 diatreme is about 96 by 128 feet and consists of kimberlite breccia; the Williams 4 is a dike-like diatreme 1216 feet long and up to 121 feet wide and consists of massive kimberlite with desultory zones of fragmental kimberlite (**Figure 39**) (Hearn and McGee, 1983).

These intrusives enclose xenoliths from the Precambrian basement, upper crust (schist, gneiss, amphibolite), lower crust (granulite, mafic granulite, amphibolite), and the upper mantle (spinel peridotite, dunite, garnet peridotite, garnet megacrysts), as well as kimberlitic indicator mineral xenocrysts. According to Hearn and McGee (1983) neither diamond nor eclogite has been found in these intrusives.

The available analyses of peridotitic garnets from the Williams kimberlites indicate the compositions are equivalent to G9. None of the garnets analyzed by Hearn and McGee (1983) fell within the G10 (sub-calcic pyropes) field. However, pressure-temperature estimates from co-existing orthopyroxene-clinopyroxene pairs in some of the peridotite nodules indicate some of the nodules may have originated from depths within the diamond stability field (Fred Barnard, written communication, 1994).

Treasure and Rosebud Counties (Porcupine Dome Field)

The Porcupine Dome Field, located in southeastern Montana, includes four intrusives known as

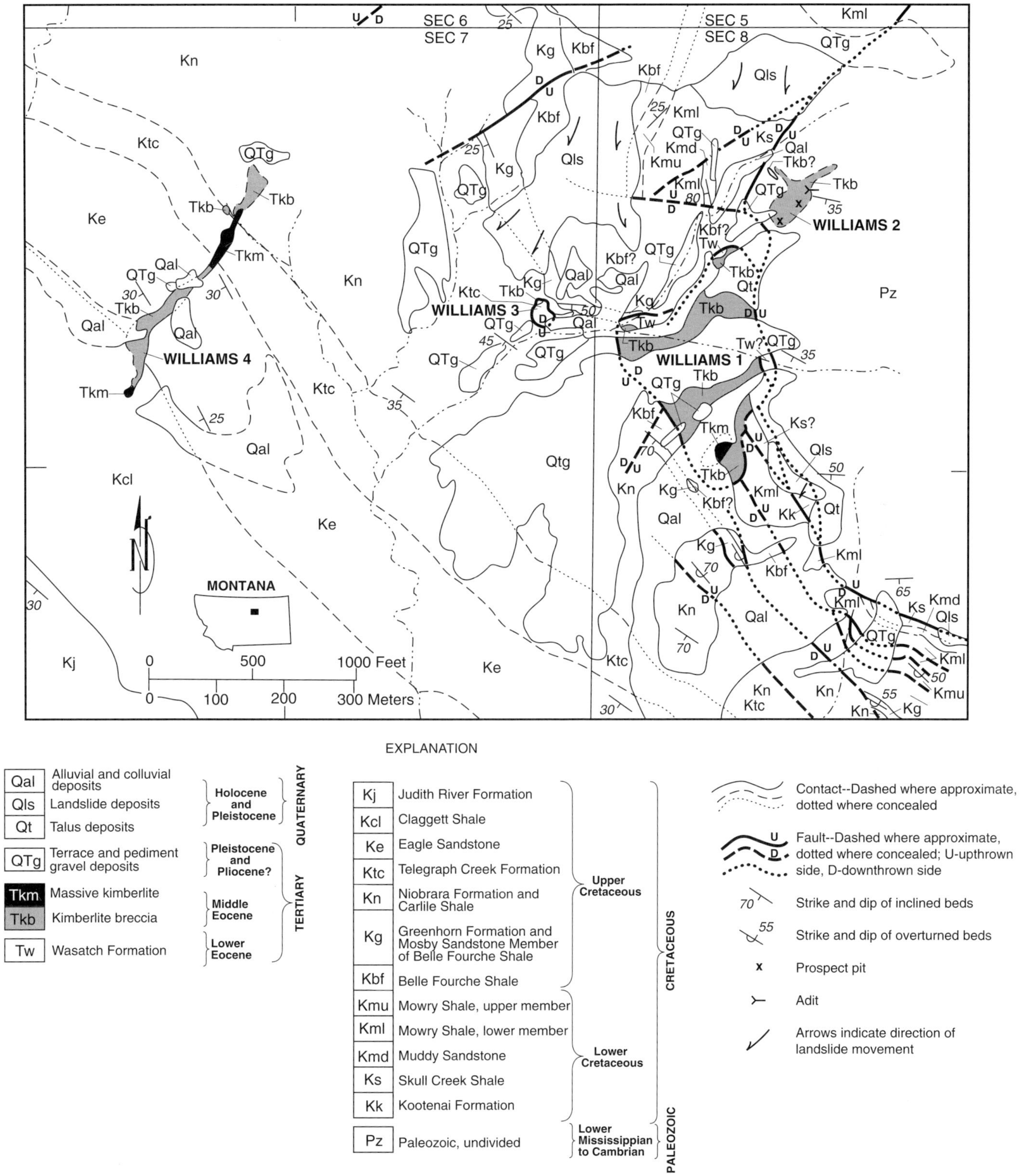

Figure 38. Geologic map of the Williams kimberlites near Zortman, north-central Montana (after Hearn and McGee, 1983).

Figure 39. Exposed hypabyssal facies Williams kimberlite in the Missouri Breaks, Montana.

Froze-to-Death Butte, Gold Butte, Geyser Spring, and Johnson Ranch Butte. Froze-to-Death Butte is a multiple vent complex located 8.5 miles northwest of Hysham (**Figure 40**). Gold Butte lies 25 miles northeast of Froze-to-Death Butte; Johnson Ranch and Geyser Spring lie between these two along strike, suggesting structural control. These intrusives are interpreted to be Eocene in age (Doden, 1997).

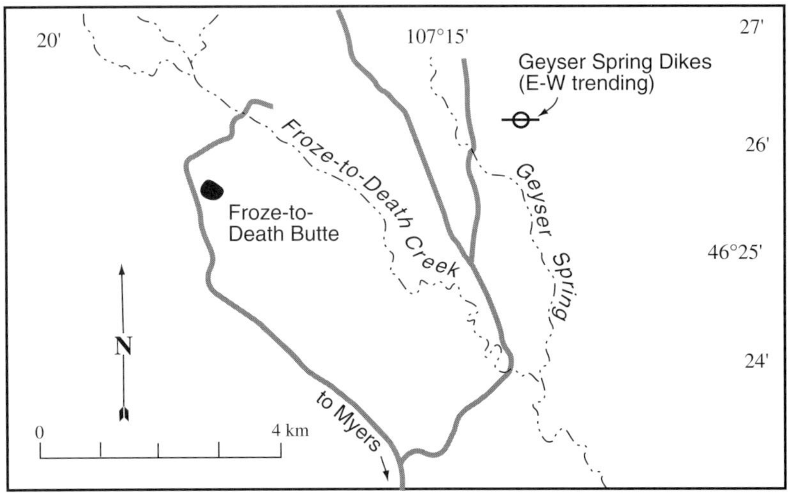

Figure 40. Location map of the Froze-to-Death Butte lamproite in central Montana (after Mitchell and Bergman, 1991).

The rocks which form Froze-to-Death Butte are described by Mitchell and Bergman (1991) as massive hypabyssal intrusive breccias consisting of altered olivine-phlogopite-diopside lamproite. In contrast, Doden (1996) indicated the rocks forming both buttes do not resemble lamproite, but are instead ultramafic lamprophyres (aillikites) that share some affinities with kimberlite. In particular, he reports the intrusives contain picroilmenite and garnet macrocrysts.

According to Doden (1996), the picroilmenites are 1 to 5 mm in size and have moderate MgO (5-13 wt.%), low Cr_2O_3 (<0.7 wt.%), and crystal rims enriched in MgO relative to the cores. These values are characteristic of kimberlitic ilmenites (Doden, 1996). Garnets (1 to 4 mm) from the Porcupine Dome area include two compositional and color ranges. One consists of pink-orange garnet that resembles eclogitic garnets. These have 0 to 0.1 wt.% Cr_2O_3, 1.0 to 4.5 wt.% CaO, 8.1 to 13.2 wt.% MgO, 22.8 to 30.8 wt.% FeO, <0.06 wt.% TiO_2, and <0.07 wt.% Na_2O. According to Doden (1996), these garnets are similar to some Missouri Breaks garnets derived from upper crustal sources.

The second type of garnet is reddish-purple and is similar to peridotitic garnets. These garnets contain 5.7 to 7.0 wt.% Cr_2O_3, 4.7 to 7.0 wt.% CaO, 18.6 to 21.6 wt.% MgO, and 5.9 to 7.9 wt.% FeO. The peridotitic garnet compositions primarily fall within the G9 compositional field; however, a few have subcalcic G10 compositions similar to garnets found as diamond inclusions (Doden and Gold, 1993).

Although the lamprophyres apparently sampled the diamond stability field, the picroilmenite compositions are depleted in Cr_2O_3, indicating that oxidizing conditions prevailed in the magma. This may suggest that diamond preservation was not favorable. However, since the magmas did apparently sample the diamond stability field, the search for undiscovered (buried) pipes of olivine lamprophyre and lamproite in this region may be productive.

Utah (Wyoming craton)

The northern part of Utah is underlain by Archean rocks of the Wyoming Province. South of the province, portions of the state are underlain by Proterozoic basement rocks. Some of these basement rocks may provide a favorable source for kimberlite or lamproite. Lamproites, minettes, and mica-peridotites are all reported in Utah, although no diamonds are known from the state.

Daggett County

Several stream sediment samples collected from streams draining to the north into the Green River Basin from the Uinta Mountains and Daggett County have yielded kimberlitic indicator minerals (McCandless and others, 1995). The source of the minerals remains to be discovered.

San Juan County

Three ultramafic diatremes are recognized south of the Wyoming craton in the Colorado Plateau of southern Utah, southeast of Mexican Hat. These are the Mule Ear, Moses Rock, and Cane Valley intrusives. The rocks were originally described as kimberlite (McGretchen and Silver, 1970), but are currently considered to be related to minettes (Mitchell, 1986).

The Moses Rock diatreme in Utah, and the nearby Garnet Ridge diatreme in Arizona, contain abundant G9 garnets with relatively low Cr_2O_3 (1 to 5 wt.%). A temperature histogram based on the nickel thermometer for the garnets shows all the garnets were derived from shallow depths within the graphite stability field (Griffin and Ryan, 1993).

Moses Rock dike. The Moses Rock dike is exposed over a 3.8-mile strike length and reaches a maximum width of 950 feet. The dike is brecciated and intrudes Permian sandstones and siltstones of the Cutler Formation. The dike has been dated at 30 Ma (Brookins, 1970a).

A detailed investigation by McGretchin (1968) suggested that the kimberlitic minerals in these breccias were derived by mechanical disaggregation of mantle rocks (spinel peridotite and garnet peridotite) from a depth of 30 to 180 miles. The breccia includes a variety of xenoliths from the surrounding country rock, as well as crustally derived igneous and metamorphic rocks, and a small number of eclogite, pyroxenite, and rare peridotite nodules (Meyer, 1976).

Cane Valley intrusive. The Cane Valley intrusive is a collapse structure about 0.5-mile in length that reportedly contains kimberlitic and carbonatitic dikes.

Summit County

Lamproites and mica peridotites are found 40 miles east of Salt Lake City near Kamas. The Moon Canyon lamproites (40 Ma), located a short distance southeast of Kamas, consist of hypabyssal olivine-sanidine-diopside-richterite-phlogopite lamproite flows and sills (**Figures 41** and **42**).

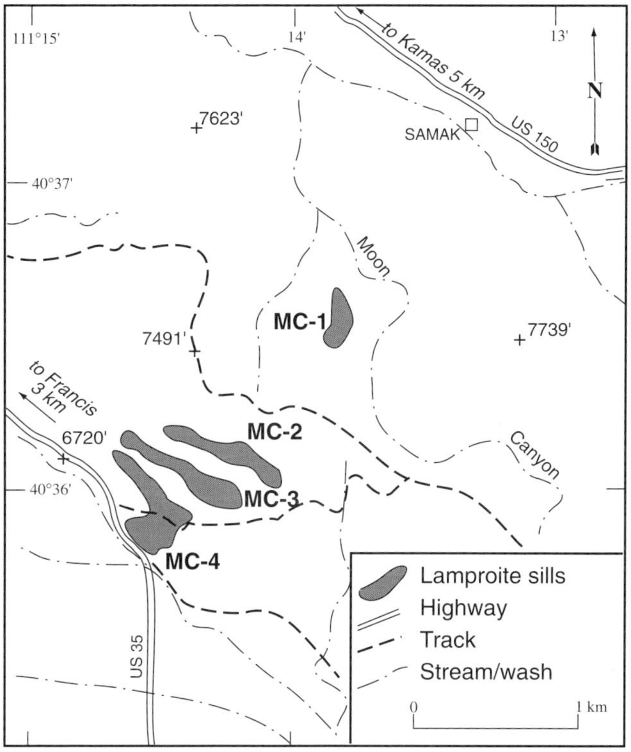

Figure 41. Location map of Moon Canyon lamproite flows and sills near Francis, Utah (after Mitchell and Bergman, 1991).

Figure 42. Flow banding in Moon Canyon lamproite, Utah.

The Whites Creek lamproites (13 Ma) to the northeast of Kamas (**Figure 43**) include a series of small orendite dikes, which have penetrated upper Cretaceous shales, 2 miles northwest of the Uinta North Flank fault zone. These lamproites contain chrome spinel, phlogopite, and olivine. Compared to the Moon Canyon lamproites, the Whites Creek lamproites have less SiO_2 and nearly twice the MgO content.

In the same general area, mica-peridotite was intersected in the Silver King tunnel in the Park City mining district. Another mica-peridotite was reported along the Erickson Basin fault in Smith-Moorehouse Canyon of northeastern Utah (Henage, 1972). The latter peridotite (39 Ma) contains melilite and biotite, suggesting the rock is possibly an alnöite.

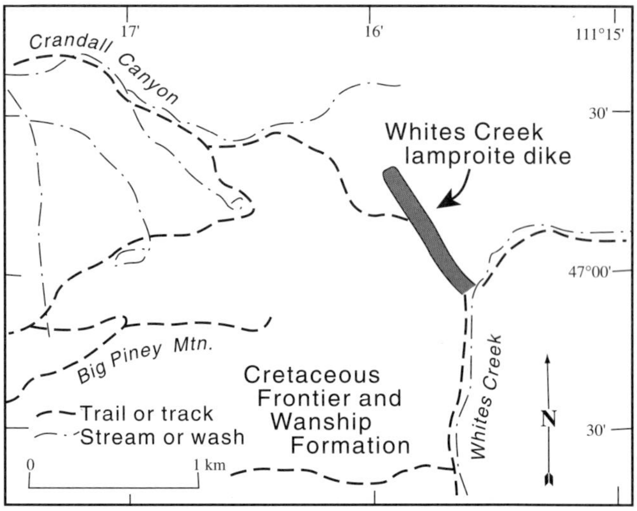

Figure 43. Location map of the Whites Creek lamproite dikes, Utah (after Mitchell and Bergman, 1991).

Great Basin–Rocky Mountain region

The Great Basin –Rocky Mountain region of the western United States includes all or portions of Arizona, Colorado, Idaho, Montana, Nevada, New Mexico, Utah, and Wyoming. Parts of this region are underlain by rocks of the Wyoming craton, which form part of an archon and proton (**Figure 5**) that is considered to be one of the more favorable terranes in the United States for the discovery of diamondiferous kimberlite and lamproite. The states of Colorado, Montana, Utah, and Wyoming have been discussed on pages 9-48 under the Wyoming craton. Thus, this section will be confined to those states outside the Wyoming craton but still within the Great Basin–Rocky Mountain region.

Arizona

Other than diamonds of extraterrestrial origin found in the Diablo Canyon meteorite east of Flagstaff (Blank, 1935), the only other reported diamonds in Arizona include a one-carat, brown diamond found near Philadelphilos, and diamonds reportedly found in the Santa Maria River several miles west of Prescott (Mather, 1941). The exact location of the Santa Maria diamonds was not given, and their authenticity is considered suspect.

Ultramafic rocks in Arizona include 30-Ma ultramafic lamprophyres in the Navajo volcanic field of the Colorado Plateau. The time of emplacement of these rocks coincided with uplift of the Colorado Plateau (Anderson and Perkins, 1975). Arizona is underlain by Proterozoic age rocks.

Apache County

A group of potassic-rich, mafic to ultramafic volcanic rocks and breccias occurs in the Navajo volcanic field of the Four Corners region of northeastern Arizona, northwestern New Mexico, southeastern Utah, and southwestern Colorado on the Colorado Plateau. Some of these rocks have similarities to kimberlite, but recent studies indicate they are probably not kimberlites, and were derived from depths too shallow for diamond.

Problems have arisen in the classification of these rocks. For example, Naeser (1971) reported the intrusives in the Navajo volcanic field (including Buell Park, Fort Defiance, Garnet Ridge, Coliseum, Hoskietso, Hopi Buttes, and Hunters Point) were minettes and monchiquites, and Roden and Smith (1979) reported the field included minette intrusives and diatremes and rare kimberlitic tuff pipes and dikes. Schmitt and others (1974) also indicated that rocks of kimberlitic affinity occur in the volcanic field. More recently, Laughlin and others (1989) indicated that the potassic-rich volcanic rocks in this region included katungites, monchiquites, olivine leucitites, vogesites, alnöites, minettes, and kimberlitic tuffs, but according to Mitchell (1986), a number of these diatremes, plugs, and dikes previously considered kimberlite (e.g., Buell Park, Mule Ear, Moses Rock, and Cane Valley) are related to minettes.

More recently, Doug Smith (written communication, 1995) reported that the ultramafic diatremes of the Navajo field are best described as: "serpentinized ultramafic microbreccias," and indicated that: "the serpentine-rich matrix of the diatremes appears to have intruded as a gas-solid mixture and no textural or chemical evidence of a melt component has been recognized in the matrix" (Roden, 1981; Smith, in press). The maximum depth sampled by these intrusives is possibly no greater than about 45 miles (Doug Smith, written communication, 1995).

The Garnet Ridge intrusive within the volcanic field consists of four small intrusives within 3 miles of one another. The largest is about 1000 feet across and consists of tuff and lapilli tuff (Meyer, 1976), and contains both chromian diopside and pyrope garnet (Allen and Bulk, 1954).

One of the intrusions in the Navajo volcanic field may have been the source of some of the pyrope garnets used in the salting of a sandstone outcrop in northwestern Colorado. Garnets from the Mules Ear area, in particular, exhibit similarities to the garnets at the diamond hoax site (McCandless and others, 1995).

Idaho

The Wyoming craton (archon) extends from Wyoming into eastern Idaho, which marks the western boundary of the craton (Hausel and others, 1991). Areas to the west of eastern Idaho are generally considered unfavorable for the emplacement of diamondiferous kimberlite and lamproite, even though a few placer diamonds have been reported in western Idaho.

For example, Kunz (1885) reported a few small diamonds had been found in placers in Idaho; according to Shannon (1926), the only authenticated diamonds in Idaho were derived from the Rock Flat gold mine in Adams County. The mine was located at the head of Little Goose Creek Canyon five miles east of New Meadows near McCall. This area lies along the west-central border of the state in an area considered unfavorable for *in situ* diamonds.

Microscopic diamonds were also reported in the black sands of the Snake River. Unfortunately, no exact location was given for these diamonds as the Snake River cuts across the entire state from east to west. The diamonds were reported to be grayish-white and more or less opaque, with a few greenish stones. Most of the crystals were described as fragments of larger diamonds (Blank, 1934), suggesting a distal source.

In recent years, reports of lamproites in the McCall area have led to some exploration for diamonds in western Idaho. However, the presence of lamproite in this region remains to be verified.

Adams County

Three small diamonds were reportedly found in 1913 in Adams County, western Idaho. The diamonds were recovered from the heavy concentrates of several cubic meters of gravel. The largest diamond was a grayish octahedron with greasy luster that weighed 0.33 carat. Ilmenite, chromite, zircon, magnetite, garnet, monazite, and corundum (ruby) were also reportedly found in the concentrates (Shannon, 1926; Sinkankas, 1959). An unconfirmed 1947 report indicated that a 19.5-carat diamond had been found at Rock Flat in the same general area, 4 miles west of McCall (Sinkankas, 1959). This diamond (?) was found in a drainage in the area of section 35, T19N, R2E, and section 2, T18N, R2E.

According to the Northern Miner (January 9, 1995), drilling in this area by Golconda Resources intersected lamproitic tuff containing pyrope garnet, chromian diopside, and ilmenite. The author was unable to verify the report.

Owyhee County

Diamonds were reported in gold placers along the Owyhee River in the southwestern corner of the state (Blank, 1934).

New Mexico

Much of New Mexico is underlain by Proterozoic basement, but only one report of a diamond is known. Due to the lack of details, it is not known if the diamond was verified.

Some mantle material is found as nodules and cognate xenocrysts in breccia pipes in the Navajo volcanic field in the Four Corners region, northwestern New Mexico. Dunite and peridotite mantle-derived xenoliths are also reported in basaltic breccia in southern New Mexico.

A large part of New Mexico, like Arizona and parts of southern Utah and Colorado, have relatively high geothermal gradients considered unfavorable for diamond deposits.

Colfax County

A lamproite plug was reportedly discovered southeast of Raton in northeastern New Mexico in the early 1980s by an exploration company searching for diamonds. No diamonds were found.

Doña Ana County

Kilbourne Hole, a late Pleistocene volcanic maar located 20 miles southwest of Las Cruces in southern New Mexico, forms a prominent crater of basalt, basalt breccia, and pyroclastics. The maar contains mantle xenoliths, and lower and upper crustal nodules including dunite and peridotite. Fuhrbach (1992) reported that some of the nodules contained gem-quality peridot, but no diamonds are reported in the area.

San Juan County

The Greens Knob intrusive in the northwestern corner of the state consists of lapilli tuff and is about 0.5 mile in diameter (Meyer, 1976). The rock was initially classified as kimberlite (D. Smith, 1979), but more recently has been recognized as a minette (Mitchell, 1986). Minettes have also been identified at nearby Fluted Rock, Outlet Neck, the Beast, Beelzebub, and Black Rock diatremes (Allen and Bulk, 1954).

Santa Fe County

A diamond was reported from gravels near Santa Fe in north-central New Mexico (Blank, 1935).

Appalachian Mountains, eastern United States

Many diamonds have been reported in the Appalachian Mountains region of the eastern United States. Diamonds have been reported from Alabama, Georgia, Kentucky, Maryland, New York, North Carolina, South Carolina, Tennessee, Virginia, and West Virginia. The authenticated diamonds range in weight from 0.25 carat to 34.46 carats. There have also been reports of diamonds weighing more than 100 carats from this region, although these have not been verified. The great majority of the stones found in the Appalachians were recovered as a by-product of historical placer gold mining.

The source of the diamonds in this region is unknown, and since much of the Appalachian region is part of the Grenville terrane which is a tecton, this region is considered unfavorable for diamondiferous kimberlite and lamproite, even though diamonds have been reported from alnöite at Ile Bizard, Quebec, Canada (Raeside and Helmstaedt, 1982) and reportedly found in peridotite in New York and Maryland. Several writers have suggested that the diamonds in the Appalachians originated from a distant source and were transported to the region by streams, glaciers, and even migrating birds from South America, or were the product of fraud. Some early writers even speculated that the diamonds were derived from nearby outcrops of itacolumite (micaceous sandstone) (Hausel and Bond, 1994). However, kimberlite, ophiolite, micaceous peridotite, and diamondiferous (?) peridotite have all been reported in the Appalachians; thus, it is possible that the source of many of the diamonds in this region originated from nearby mantle-derived host rocks.

Many of the kimberlites found in the Appalachians are reportedly localized at the intersections of reactivated faults with cross-structural lineaments (Parrish, 1984). Such lineaments may have provided access for several undiscovered kimberlites.

Alabama

Alabama forms part of the coastal plain of the United States, and is not considered to be a favorable terrane for diamonds, except for the occasional stones that have been stream transported from host rocks outside Alabama. However, diamonds have been reported from three localities in Alabama (Sinkankas, 1959), apparently all found in placers. These stones, ranging from 2.41 to 4.61 carats, are documented, and two of the stones were donated to the American Museum of Natural History. Two of the diamonds were found near the southern margin of the Appalachian uplift.

Lee County

A distorted 4.61-carat octahedron was found in Lee County along the eastern margin of the state near the Chattahoochee River. The diamond was placed in the American Museum of Natural History (Blank, 1935; Sinkankas, 1959).

St. Clair County

In 1905, a greenish, 2.41-carat diamond was found near Prescott Siding about 1.25 miles east of Brompton in the northeastern portion of the state (Blank, 1935).

Shelby County

In 1900, a faint yellow octahedron was found near Birmingham in the northeastern portion of the state. The diamond was placed in the American Museum of Natural History (Sinkankas, 1959) and weighed 4.37 carats (Blank, 1935; Mather, 1941).

Georgia

Georgia is considered unfavorable for *in situ* diamonds in that much of the state is located within the coastal plains. Even though the Appalachian uplift lies along the northwestern margin of the state, the Appalachians are part of the Grenville tecton, which is an unfavorable terrane for diamondiferous kimberlite and lamproite. However, the available historical reports indicate that possibly many diamonds may have been found in placer gold mines in Georgia.

The source of the diamonds is unknown, and most of the discoveries are poorly documented, so the actual number of diamonds found could have ranged from several to dozens. The great majority of the stones were reported from the Appalachian region in the northern portion of the state (**Figure 1**).

The diamonds were described as ranging from microdiamonds to possibly the largest diamond ever found in the United States (although this stone was not verified). According to Cook (1978), diamonds have been reported from Banks, Bartow, Burke, Camden, Carroll, Cherokee, Clayton, Cobb, Dawson, Forsyth, Gwinnett, Habersham, Hall, Haralson, Lee (?), Lumpkin, Paulding, Twiggs, and White Counties (**Figure 1**). Unfortunately, very few of the reported gemstones were verified. The first report of diamonds in Georgia was in 1843, when diamonds were recovered from gravels in placer gold mining operations in Hall and White Counties (Sinkankas, 1959). The reported diamond occurrences are described below.

Burke County

A hexoctahedron weighing 7.11 carats was found at the old Shell Bluff post office along the eastern edge of the state (Cook, 1978).

Camden County

Two 50-mesh diamonds were identified in the heavy concentrates of stream placer samples from Camden County along the coast. One stone was a colorless hexoctahedron (Cook, 1978).

Clayton County

In 1887, a 4.25-carat yellow octahedron (9 x 10 x 7 mm) with one dark inclusion was found on the Daniel Light farm three-quarters of a mile northeast of Marrow station (12 miles south of Atlanta) in the northern portion of the state. A second stone, possibly from Clayton County, was found in 1889 at an undisclosed location near Atlanta. This was a poor quality 2-carat stone (Cook, 1978).

Hall County

In 1843, a diamond was recovered at Winns or Williams Ferry at the mouth of a small creek that enters Muddy Creek, about a half-mile from its junction with the Chattahooche River in the northern portion of the state. The stone weighed more than 6 carats. The country rock was reported to consist of granite and pegmatite.

There is an intriguing story about diamonds being found in gravel along the Stockeneter Branch of Glade Creek, 12 miles northeast of Gainsville. The property was mined in the 1850s for gold, and diamonds were frequently encountered in the sluice box clean-ups. Several pounds of diamonds were reportedly saved by a Dr. Loyd and his family. The diamonds were reported to average 4 carats in weight and included three very large stones–the largest weighed more than 100 carats. This large stone may have been quartz.

A Mr. Stephenson also reported finding several small, well-formed stones in the area of the Glade Gold Mine several years later (Cook, 1978). The U.S. Geological Survey also reported diamonds in the vicinity of Glade Creek (U.S. Geological Survey, 1968).

Lee County

A diamond found in Lee County was sold to Tiffanys in 1901. The stone was a flattened hexoctahedron of 3.5 carats and was white with a greenish tint (Cook, 1978). The reported discovery was described to be near Columbus, Georgia, which would suggest that the diamond may have been found in Lee County, Alabama, instead of Georgia, since Columbus lies immediately east of Lee County, Alabama. Lee County, Georgia lies 50 miles southeast of Columbus.

Twiggs County

A number of diamonds were reported on the Nelson property 10.5 miles southeast of Macon, or about 1.5 miles northeast of Pikes Peak Station in the center of Georgia (Cook, 1978).

White County

Some small diamonds were reported from gold placers in the Nachoochee Valley in the northeastern corner of the state. These included a 3-grain diamond recovered from a long tom in White County.

Three small diamonds (0.125-, 0.15-, and 0.5-carat, respectively) were also found in the Horshaw gold placer in the Nachoochee Valley near Loudsville. These were described as opaque with no definite shape. Another diamond was reported from Lot 10 on the Lumsden property (the Lumsden property may be part of the Horshaw mine according to Cook, 1978). Other reports indicate stones of good quality were found in this region. For instance, a perfect diamond weighing nearly a carat was reportedly recovered from a gold mine in the county.

Kentucky

Much of Kentucky lies within the Appalachian uplift and is underlain by basement rocks of the Grenville tecton. Thus the state would be considered unfavorable for the discovery of commercial diamond deposits.

Kimberlite, mica-peridotite, and at least one diamond have been found in Kentucky. The kimberlites in eastern Kentucky have received a fair amount of exploration interest in the past, and based on the chemistry of some megacrysts and nodules from the kimberlites, it is possible that the intrusives originated within the diamond stability field. However, there are no verified reports of diamonds from the kimberlites, and bulk sampling of the Ison Creek kimberlite by Cominco American Inc. in the early 1980s yielded no diamonds.

Adair County

According to Sinkankas (1959), a yellowish 0.766-carat diamond was reported from a stream near Montpelier in southern Kentucky. This is probably the same diamond that was reported in Russell County in the same general part of the state.

Crittenden County

Approximately 20 ultramafic igneous dikes (257 Ma, age date from Zartman and others, 1967) occur near the town of Marion in western Kentucky. The dikes are part of a larger field of about 50 intrusives which extend into southern Illinois (Koenig, 1956).

These intrusives were initially described as mica-peridotite and occasionally referred to as kimberlite. However, the overall mineralogy, especially the paucity of garnet and pyroxene, is not consistent with kimberlite (Meyer, 1976). The rocks are more appropriately classified as mica-peridotite or lamprophyre depending on the presence or absence of olivine or olivine pseudomorphs.

Elliott County

Three kimberlite intrusives (two diatremes and one dike) occur in Elliott County, eastern Kentucky, along Ison Creek and Hamilton Branch (**Figure 44**). The intrusives cut Lower Pennsylvanian sedimentary rocks and have been dated at 279 and 270 Ma (Permian) (Zartman and others, 1967). Aeromagnetic surveys over the region identified a 200-gamma magnetic anomaly associated with the kimberlites (Parrish and Lavin, 1982).

The two diatremes consist of multiple phases of diatreme and hypabyssal facies kimberlite dominated by kimberlite breccia with less common massive kimberlite. Sparse stream sediment sampling in the area suggests the intrusives are either more

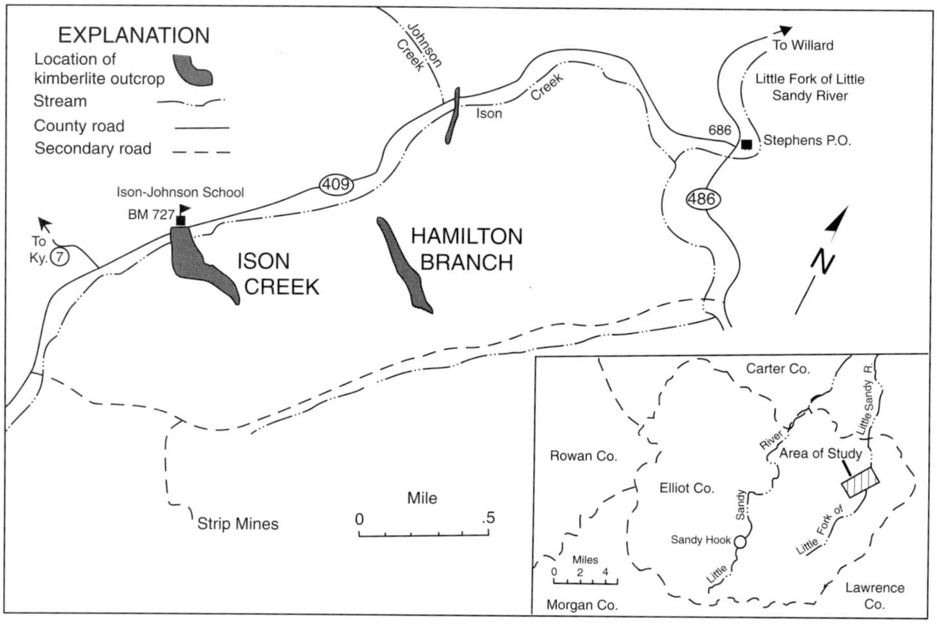

Figure 44. Location map of the Ison Creek and Hamilton Branch kimberlites, Kentucky (after Bolivar, 1982).

widespread than previously reported, or that an undiscovered intrusive (or intrusives) occurs nearby (J.E. Bond, personal communication, 1994).

The intrusives consist of phlogopite-poor, ilmenite- and pyrope-rich kimberlite containing some mantle and crustal xenoliths, xenocrysts, and megacrysts (Schulze, 1985). The host kimberlite is inequigranular with common phenocrysts and metacrysts of serpentinized olivine, pyrope garnet, picroilmenite, phlogopite, and minor pyroxene. Studies indicate the ilmenite contains appreciable magnesium (Meyer, 1976). One published ilmenite analysis yielded 9.39 wt.% MgO and 0.21 wt.% Cr_2O_3 (Garrison and Taylor, 1980). The groundmass is serpentinized and carbonatized and consists of serpentine and calcite with accessory perovskite, apatite, and chlorite (Bolivar, 1982).

Equilibration temperatures and pressures determined for some metacrysts and garnet lherzolite nodules from the Kentucky kimberlites fall within the diamond stability field (Taylor, 1984). In other words, the chemistry of these metacrysts and nodules suggest these intrusives may have originated within an area of the mantle where diamonds are stabile. Garnets of peridotitic paragenesis are predominately calcic-chrome pyropes of lherzolitic affinity; however, a few sub-calcic, chrome pyropes have also been recovered from the area.

The intrusives were tested for diamond content shortly after their discovery in 1885, and a 72-foot-deep shaft was sunk on the southernmost intrusive. Later exploration occurred in 1907, when the Kentucky Kimberlite Diamond Mining Company was established in Catlettsburg, Kentucky. Between 1965 and 1970, Kentanna Minerals from Henderson County, Kentucky completed additional trenching and a washing operation tested colluvium near the southern end of the southeastern-most intrusive (Brown, 1977) (**Figure 45**). At least one of the kimberlites was also tested by Cominco American Incorporated and Superior Minerals Company in the early 1980s; Janse (1994b) reported it to be barren. However, local residents report that three diamonds were recovered by the Kentucky Kimberlite Diamond Mining Company, and another was reportedly recovered from a nearby drainage and cut into a gemstone (J.E. Bond, personal communication, 1994). The lack of common G10 pyrope garnets from this kimberlite, however, suggests that the intrusive may be a poor exploration target.

Russell County

There is at least one verified report of a diamond in Kentucky. This diamond was recovered from Cabin Fort Creek, Russell County, in the southern portion of the state. The diamond weighed 0.776 carat and was incorporated into the U.S. National Museum's collection (Holden, 1944). There are no known mafic or ultramafic intrusives near this site.

Maryland

Maryland lies along the east coast of the United States adjacent to the Grenville tecton. The state is considered unfavorable for diamondiferous host rocks based on current exploration concepts; however, both kimberlite and diamondiferous peridotite have been reported in Maryland. The report of diamondiferous peridotite in Maryland is intriguing, in that no recent reports could be found on this occurrence.

a. b.

Figure 45. (a) Exposed Ison Creek kimberlite with (b) the nearby remains of a past diamond testing operation.

Baltimore County

Diamonds were reportedly discovered *in situ* in an altered serpentinized peridotite at Bare Hills, northern Maryland (Mather, 1941). No other information could be found on this occurrence.

Washington County

Kimberlite is reported near Clear Springs in northwestern Maryland. The Clear Springs intrusive appears to lie on a northeasterly-trending structure which also controlled the emplacement of the Mt. Horeb, Virginia and Norris, Tennessee kimberlites to the southwest (Parrish and Lavin, 1982) (**Figure 46**). This belt of kimberlite intrusives parallels a second belt of kimberlites farther to the west, which includes those in Kentucky, Pennsylvania, and New York. Both of these belts should provide exploration targets for additional kimberlites.

New York

The western portion of New York state is located within the Grenville tecton. Kimberlite in the state has been reported in the southern portion of the state in the Ithaca field south of Syracuse. Kimberlite is also reported near Syracuse and has been reported near Mannhein. Along with these intrusives, a diamond was reportedly found *in situ* in a peridotite near Syracuse. It is not known if the find was verified, nor is it known if this particular peridotite is the same intrusive as the kimberlite reported near Syracuse by Parrish and Lavin (1982). Detrital diamonds have also been reported from three other counties in the state.

The Ithaca field lies along a major trend which includes kimberlites in Pennsylvania and Kentucky to the southwest (Parrish and Lavin, 1982). The possibility of additional discoveries of kimberlite along this trend is considered good. The possibility of the kimberlites containing commercial amounts of diamond is considered unlikely.

Clinton County

Diamonds have been reported in glacial drift at Plattsburg in the northeastern corner of the state (Mather, 1941). No details were given.

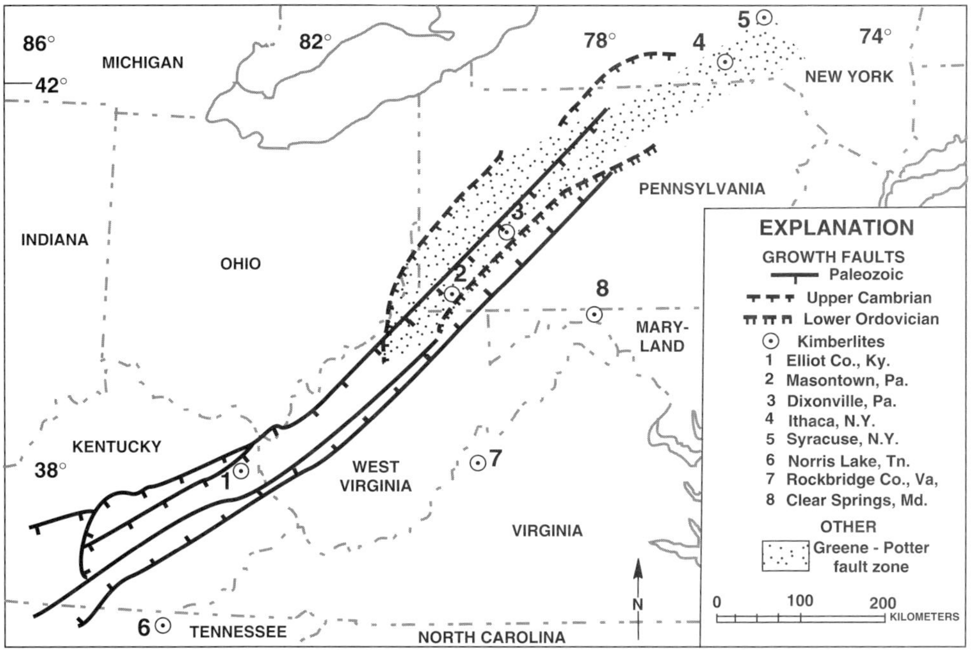

Figure 46. Kimberlite and associated lineaments in the eastern United States (after Parrish and Lavin, 1982).

Herkimer County

Kimberlite was reported near Mannhein east of Little Falls (Wagner, 1914).

Onondaga County

Blank (1934) reported that a microscopic diamond was found in peridotite near Syracuse, New York, north of Ithaca. The diamond apparently was described in 1921 after a microscopic study of the peridotite by a geologist at Syracuse University (Mather, 1941). Kimberlite is also reported near Syracuse (Parrish and Lavin, 1982). It is unknown whether or not the reported kimberlite and peridotite are the same intrusive.

Putnam County

A diamond was reportedly recovered from a butchered chicken at Cold Spring, New York, near the southeastern corner of the state (Mather, 1941). No other details were given.

Tompkins County

The Ithaca kimberlite field south of Syracuse includes several kimberlite dikes and one diatreme (**Figure 47**) (Martens, 1924; Kay and others, 1983). The exposures are described as narrow dikes of kimberlitic affinity that range from a few inches to 17 feet wide, and are found in stream and road cuts in folded Devonian sedimentary rocks (Martens, 1924). Emplacement of the intrusives was controlled by a prominent set of north-south trending joints.

More recent work by Kay and others (1983) expanded the Ithaca field to 82 dikes and one small diatreme known as the Poyer Orchard diatreme. Kay and others (1983) examined three of the intrusives and suggested the kimberlite magmas originated from depths of less than 90 miles. Both G9 pyropes and chromian diopside were recovered from the kimberlites.

The dikes range from relatively fresh kimberlite with visible red garnets and chromian diopside to highly altered rock (Kay and others, 1983). The altered dikes consist of mica and serpentine with rare phenocrysts of olivine. The groundmass consists of perovskite, apatite, and magnetite with ubiquitous serpentine, chlorite, and calcite. Martens (1924) recorded the presence of small amounts of chromite, picotite (chromian spinel), graphite, red garnet, and bright green diopside.

The presence of melilite in some of the dikes led Martens (1924) to refer to these rocks as alnöites. However, later work by Kay and others (1983) disproved the presence of melilite. Foster and Reitan (1972) suggested that most of the dikes are kimberlitic based on texture, composition, and mineralogy. Three of the intrusives examined by Kay and others (1983) were reported to contain mantle xenocrysts and rare xenoliths. Clinopyroxenes of mantle origin from the kimberlites yielded compositions >0.5 wt.% Cr_2O_3 and 0.5 to 2.0 wt.% Na_2O (Kay and others, 1983).

St. Lawrence County

A diamond was reported by Sinkankas (1959) in northern New York from the Grass River near Massena. The report is unconfirmed.

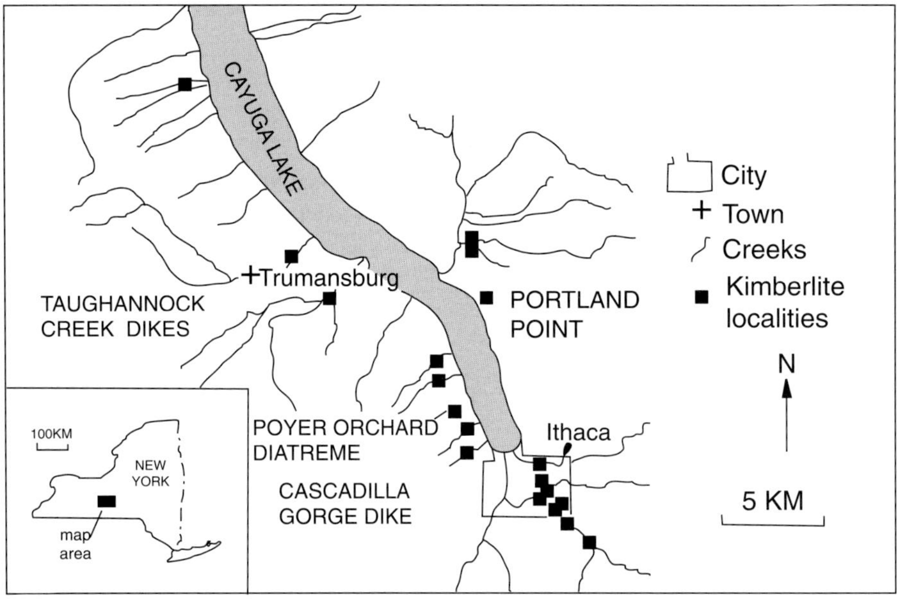

Figure 47. Location map of the Ithaca, New York kimberlite field (after Kay and others, 1983).

North Carolina

North Carolina lies along the coastal plain of the east coast of the United States, and is bordered on the west by the Appalachian Orogen and Upper Piedmont. Diamond, chromite, pyrope, and menaccanite [according to Sinkankas (1964), menaccanite is ferroan geikeilite (picroilmenite?)] have been reported at several localities in North Carolina, suggesting that kimberlite, or a similar undiscovered intrusive, may occur within the state.

North Carolina has also produced several diamonds, some of probable gem quality. Several of the diamonds have been found in the Upper Piedmont region of North Carolina, which is underlain by Precambrian schists (Late Proterozoic) and includes a peridotite belt west of the known diamond localities (Blank, 1934). Whether or not the peridotites are host to mantle material is unknown.

Diamonds were first discovered in North Carolina in 1843 associated with placer gold mining activities, and other diamonds were periodically found until 1893. The exhaustion of auriferous gravels, and the sharp reduction of placer gold mining, led to the decline of the diamond discoveries (Sinkankas, 1959).

Burke County

In 1843, an octahedral diamond was discovered at the ford of Briddle Creek in the Appalachian region, western North Carolina. This was followed by the discovery of a second diamond nearby. In total, at least three diamonds have been reported near Briddletown (J.E. Bond, written communication, 1987).

Diamonds ranging from 0.5 carat to over 2.0 carats were reported at the J.C. Mill gold mine. According to Blank (1934), the Mill diamonds were all questionable and may have instead been quartz or zircon. Kunz (1907), however, reported one of the diamonds to be an octahedron (unlike quartz or zircon!). Diamonds were also found nearby at Dyortsville in McDowell County, about 4 miles southwest of Briddletown. A diamond (0.31 carat) from this region was incorporated in the State Museum at Raleigh (Sinkankas, 1959).

It may be significant that ilmenite and chromite were also reported from the Mill gold mine. Pyrope garnet has also been reported in the gold washings of Burke County (Kunz, 1885), and ilmenite was reported from the Linnville Mountain area (J.E. Bond, written communication, 1987).

Cleveland County

Blank (1934) reported that a small, bright, canary yellow stone weighing 0.75 carat was found near Kings Mountain in southwestern North Carolina, in 1893.

Franklin County

Two diamonds were found at the Portis mine in the northeastern portion of the state. One of the diamonds was described as a beautiful octahedron (Blank, 1934; Kunz, 1885).

Lincoln County

A greenish, elongated hexoctahedron, approximately 0.5 carat, was found in 1852 near Cottage Home in the southwestern portion of the state. This diamond was transparent but had a greenish hue (Sinkankas, 1959).

Madison County

Kunz (1885) reported that purple almandines to red pyrope garnets were found four miles from

Marshall in western North Carolina. [Author's note: almandine is generally red to reddish-brown and pyrope is typically purplish-red, lavender or yellow orange. Kunz may have reversed the descriptions of the garnets.]

Chromite and menaccanite (ilmenite) were reported from Carter's mine, and ilmenite was also reported from the Big Laurel and Haynie mines (Genth, 1891).

McDowell County

Two or three small diamonds were discovered in the headwaters of Muddy Creek in western North Carolina (Blank, 1934). Sinkankas (1959) also reported that several diamonds were found at this locality.

The three diamonds were recovered from a placer gold mine at Dysortville (U.S. Geological Survey, 1968). One was a "fine stone" that was picked up at Dysortville on the Bright farm in 1886. This specimen was a distorted, twinned, 4.33-carat hexoctahedron, reported to be transparent with a grayish-green tint. The diamond was placed in the Tiffany-Morgan collection of the American Museum of Natural History (Blank, 1934). In 1877, another stone was reported from this region that weighed 2.38 carats. This diamond was flawed, white, irregular and flattened (Blank, 1934). Pyrope and ilmenite (menaccanite) have also been reported in the gold washings of McDowell County (Genth, 1891).

Mecklenburg County

In 1852, a perfect, white, one carat stone was found at Todd's Branch in southwestern North Carolina. A "beautiful stone as large as a chinquapin" was also found here (Holden, 1944) [author's note: a chinquapin is a small nut]. According to Sinkankas (1959), a black crystal of good size was also reported from this locality, and was smashed by a hammer in the mistaken belief that diamond was indestructible.

Rutherford County

A distorted, yellow, clear and flawless 1.33-carat hexoctahedron, with curved faces was found in 1845 in the gold washings at Twitty's mine in western North Carolina (Blank, 1934). At about the same time, another diamond was reported from this locality (Sinkankas, 1959). Nearby, a 0.835-carat stone was found in a placer mine on the Levinthorpe property. The Levinthorpe diamond was incorporated into the U.S. National Museum (Holden, 1944). Another poor quality diamond was also described from the Levinthorpe placer by Kunz (1885). Ilmenite (menaccanite) is also reported in the gold placers of the county (Genth, 1891). There is another report of a diamond found in Cane Creek (U.S. Geological Survey, 1968).

Warren County

Pyrope was reported in the gold washings of Warren County in northern North Carolina. The pyrope locality lies north of the two reported diamonds discovered in Franklin County.

Pennsylvania

Portions of Pennsylvania lie within the Appalachian orogen, and several mica-peridotite dikes were described from the southwestern part of the state in the 1950s. At least two of these peridotites were reclassified in the 1980s as kimberlite. Elsewhere in the state, some pyrope garnets have been reported; however, no diamonds are known.

According to Koenig (1956), several peridotite dikes occur in the southwestern part of the state that intrude Pennsylvanian age sedimentary rocks. Within this same region, two kimberlite intrusives have been recognized–one at Masontown and the other at Dixonville (Parrish and Lavin, 1982).

A gravity anomaly, known as the Vozoff gravity high occurs in north-central part of Pennsylvania east of the known kimberlites. Parrish (1984) suggested that this anomaly may be related to kimberlite magmatism.

Chester County

Kunz (1885) reported that a purplish-red garnet with "sharp angles" was found in the Avondale quarry, Chester County, southeastern Pennsylvania. The color is suggestive of pyrope; however, the sharp angles suggest the garnet had dodecahedral habit, and probably was not a mantle megacryst, as mantle megacrysts are typically rounded due to dissolution effects produced during transportation by the magma.

Delaware County

A dark red variety of garnet, similar to pyrope in color, was found in Darby Creek near Lazaretto in

Delaware County, southeastern Pennsylvania. This occurrence was east of the Chester County occurrence (Kunz, 1885).

Fayette County

A mica-peridotite dike was recognized by Kemp and Ross (1907) near Masontown, Fayette County, southwestern Pennsylvania. This peridotite was reclassified as kimberlite by Parrish and Lavin (1982) and Parrish (1984), and according to Taylor (1984) contains a well-preserved megacryst suite typical of kimberlite.

The kimberlite dike intrudes a coal seam along a northwesterly-trending fault. Sosman (1938) concluded that the dike had an emplacement temperature of approximately 500°C based on the coking effects of the coal. Geophysical studies show the kimberlite to be associated with a weak (100 gamma) magnetic anomaly (Parrish and Lavin, 1982).

Indiana County

A second mica-peridotite dike similar to the one in Fayette County is reported near Dixonville in Indiana County, 35 miles to the north. Parrish and Lavin (1982) and Parrish (1984) reclassified the rock as kimberlite.

This kimberlite varies from 1 to 50 feet wide, and intrudes Pennsylvanian-age rocks, including coal. The rock has porphyritic texture with phlogopite, calcite, and ilmenite phenocrysts (Honess and Graeber, 1924). Olivine is not abundant, but when present shows varying degrees of serpentinization. Perovskite, green diopsidic pyroxene, titanium magnetite, spinel, and garnet all occur in minor amounts (Meyer, 1976). The contacts between the intrusive and coal are sharp and show little evidence of metamorphism (Deines, 1968). An apparent K-Ar date of 185 Ma is consistent with the geology (Meyer, 1976).

Rhode Island

Peridotite is reported near Cumberland in Rhode Island (Blank, 1935). There are no known reports of diamond in the state.

South Carolina

According to Sinkankas (1959), unsubstantiated reports indicated diamonds were found in gold placers in some of the state's northern counties. Another report indicated that a white diamond was found in a South Carolina placer, but no specifics were given.

Some vermiculite deposits in the Enoree district, west of the Kings Mountain belt (see also Cleveland County, North Carolina), were examined by Bergman (1987) and interpreted as metamorphosed lamproites. Mineralogically, these rocks consist of phlogopite, diopside, tremolite (or talc), K-feldspar, apatite, sphene, monazite, and zircon. If these were originally lamproite, only a few minerals characteristic of lamproite have been preserved.

Although the mineralogy was modified by metamorphism, the geochemical characteristics of these rocks are nearly identical to the average lamproite (Bergman, 1987). If these represent altered lamproite, possibly some vermiculite deposits in this region could provide diamond exploration targets as long as diamond was initially present and was not destroyed during metamorphism. Bergman (1987) for example, reported that Late Proterozoic to Early Paleozoic vermiculite deposits occur throughout the Appalachian Mountains and Piedmont.

Tennessee

The Appalachian Mountains continue through Tennessee, although much of the terrane is underlain by Late Proterozoic rocks of the Grenville tecton.

Diamonds and kimberlite have both been reported in Tennessee. In 1956, Koenig reported that two "good quality" diamonds had been found in 1904 in gravels immediately downstream from the Norris kimberlites in eastern Tennessee. Kunz (1907) reported that three other diamonds were found on Koko Creek, at the headwaters of the Tellico River in eastern Tennessee. The exact locations of these diamonds were not given, but were reported to have been derived from the bench lands of the Smoky or Unaka Mountains.

Monroe County

Several diamonds were reported from Monroe County about 50 to 55 miles south of the Norris kimberlite in Union County, southeastern Tennessee (Holden, 1944).

Roane County

A diamond was reported at Union Crossroads in eastern Tennessee, about 30 to 35 miles downstream (southwest) from the Norris kimberlites

(Holden, 1944). According to Sinkankas (1959), the diamond was found on the south bank of the Clinch River, and weighed 3 carats. The stone was later cut into a 1.25-carat gem.

Union County

Two kimberlite intrusives, known as the Norris kimberlites, crop out along the northern shore of Norris Lake reservoir, 25 miles north of Knoxville in northeastern Tennessee (**Figure 48**). One of the intrusives lies under the reservoir, and is only exposed during the winter when the reservoir is low. The second intrusive is located primarily above the water line with a small portion extending into the reservoir (**Figure 49**). The intrusives were first recognized in 1869 and were classified as mica-peridotite in 1927 (Meyer, 1976). Later work by Meyer (1975) showed the peridotites to be kimberlite. The kimberlites occur as distinctly foliated serpentine mica schist with no apparent megacrysts (**Figure 50**). However, numerous tiny pyrope garnets occur in a drainage cutting the schist, many of which are calcic pyropes and probably were derived from shallower depths than diamonds.

The intrusives trend northeasterly for nearly 3000 feet. Based on aeromagnetic data, they are interpreted as near-vertical plugs intruding Silurian and Devonian strata, and some nearby geophysical anomalies may indicate the presence of additional kimberlites (Tim Neal, personal communication, 1994). The rock is heavily weathered and altered, and consists of mica, hematite, ilmenite, and serpentine. Phlogopite is abundant and altered to vermiculite, and carbonate is ubiquitous. Ilmenite and red and orange pyrope garnet occur as uncommon rounded xenocrysts. The ilmenite is picroilmenite and contains approximately 8 to 15 wt.% MgO (Taylor, 1984). Xenoliths recovered from the intrusives include crinoid-bearing limestones and a garnet-pyroxenite nodule. Using geologic relationships, Meyer (1976) reported the intrusives as early Mississippian to Permian.

A shaft was once sunk into one of the intrusives. The results of the venture are unknown; however, two alluvial diamonds were reportedly found near the kimberlites. A Knoxville jeweler purchased a 3-carat diamond in 1889 and a second diamond in 1900. The second diamond was described as a white, flawless, 1.69-carat stone [author's note: Sinkankas (1959) reported this diamond weighed 1.81 carats]. The stone was found on the bank of Flat Creek near Hickory Creek Shoals about 2.5 to 3 miles downstream from the kimberlites (Holden, 1944; Hall and Amick, 1944). Testing by Cominco American Incorporated on at least one of the kimberlites in the early 1980s did not find any diamonds (Janse, 1994b).

Virginia

Virginia also lies within the coastal plain of the eastern United States, with the western part of the state lying within the Appalachian orogen. Kimberlite, diamonds, mica-peridotite, and other ultramafic rocks have been reported in Virginia. One of the largest stones recovered in the United States, the Dewey diamond (23.75 carats), was found in the James River valley.

Augusta County

A large number of igneous dikes occur within the Paleozoic sedimentary sequence of the Appalachians in Augusta County near Staunton, northwestern Virginia. These dikes have alkalic affinity and include teschenites, camptonites, and nepheline syenites. Johnson and Milton (1955) also describe mica-peridotite from this area. The dikes appear to be associated with northwest-trending joints, and are considered to be Jurassic. According to Meyer (1976), the mica-peridotites had not been adequately characterized, but Parrish (1984) later reported kimberlite in the Staunton area.

Henrico County

Diamonds from Virginia include the famous Dewey diamond found in 1885 near Manchester (Richmond) in eastern Virginia. The site of the diamond discovery is now part of the city of Richmond (Sweet, 1997). The diamond consisted of a slightly rounded, trigonal, trisoctahedron weighing 23.75 carats; it was cut into an 11.69-carat gemstone. The stone had a faint greenish-white color and perfect transparency (Kunz, 1885).

The source of the diamond is unknown, but was believed to have been transported by the James River from the Virginia gold fields during spring flooding (Blank, 1934). Potential source rocks lie a considerable distance to the west in the Appalachian Mountains, and the diamond could have been derived from a region that included the Mount Horeb kimberlite (Sweet, 1997).

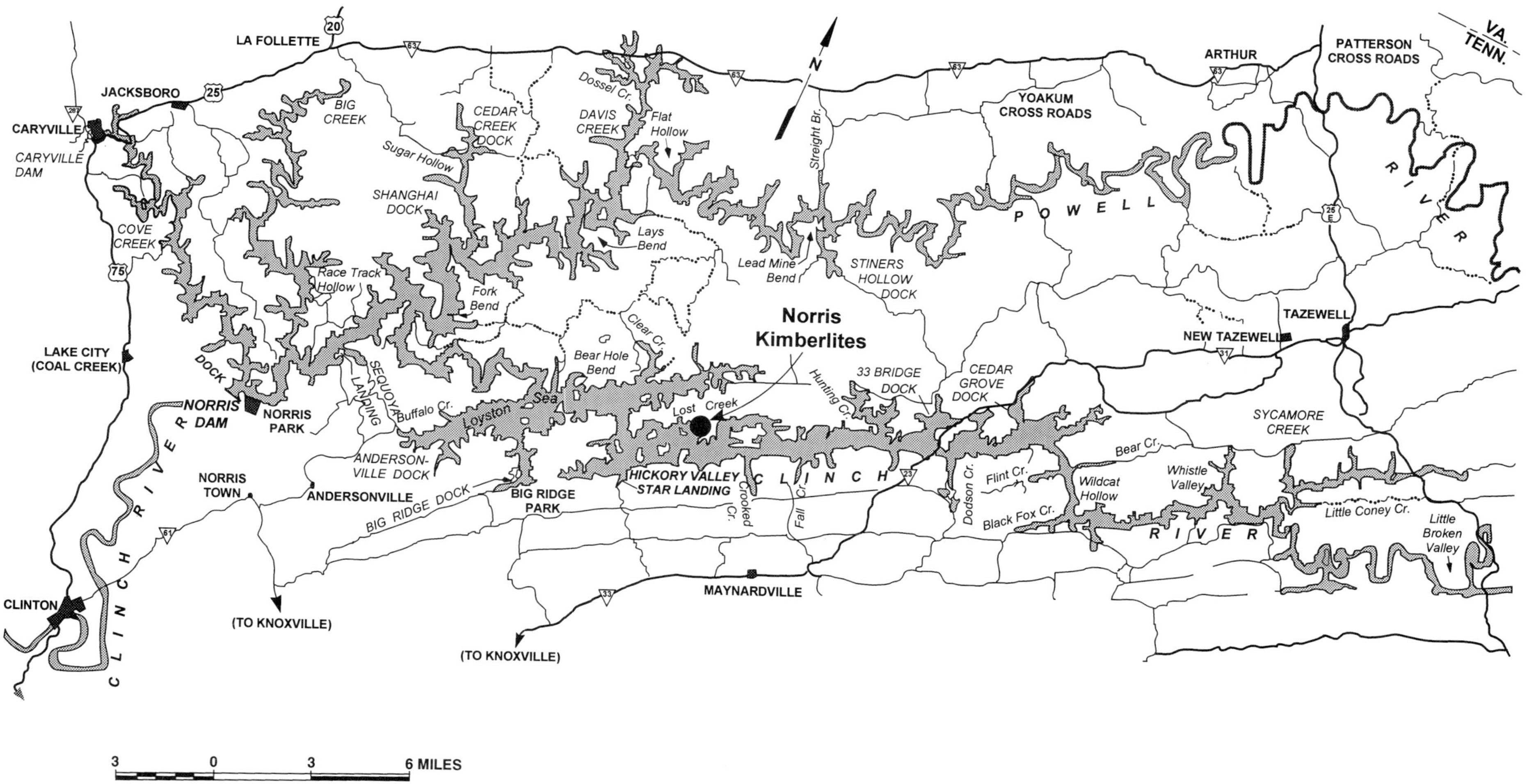

Figure 48. General map showing the location of the Norris kimberlites, Tennessee (modified from Hall and Amick, 1944).

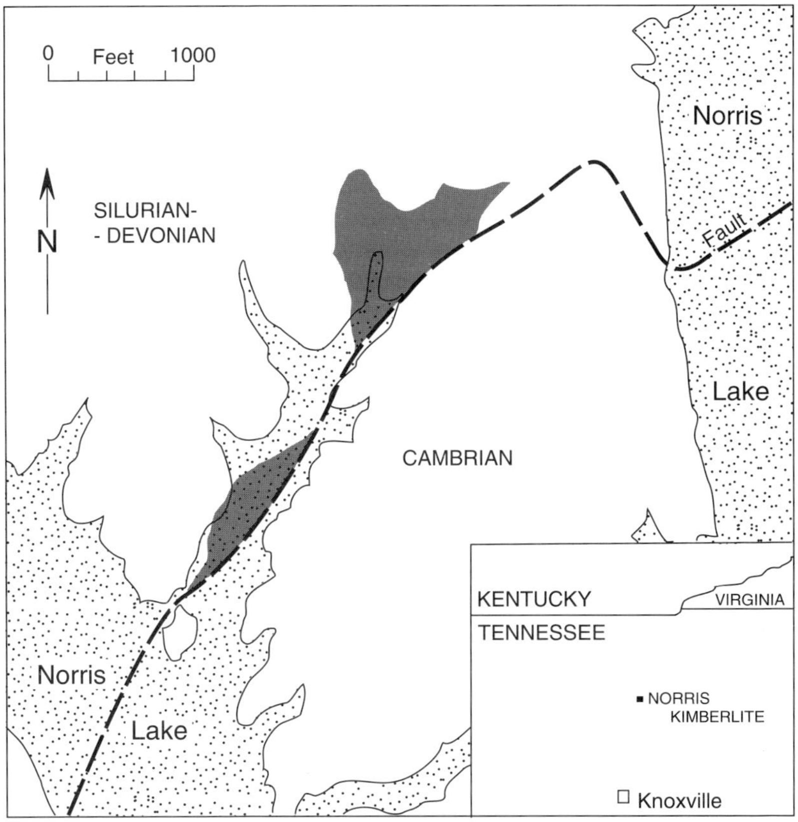

Figure 49. Location map of the Norris kimberlites (dark stippled pattern) in relation to the Norris Reservoir shoreline (after Meyer, 1976).

Figure 50. Sample of Norris kimberlite exhibiting distinct foliation.

Orange County

In 1847 a diamond was found 40 to 50 miles east of the Staunton kimberlite in the Vaucluse gold mine, north-central Virginia (Sinkankas, 1959). Details are lacking.

Rockbridge County

The Mt. Horeb intrusive, located 30 miles north of Roanoke in western Virginia, was initially classified as mica-peridotite, similar to the Staunton kimberlite in Augusta County to the northeast. This peridotite was more recently reclassified as kimberlite based on mineralogy, structure, and lack of thermal metamorphism of the sedimentary xenoliths in the host intrusive (Sears and Gilbert, 1973). The rock is altered to montmorillonite, vermiculite, and chlorite, and contains some chromian-spinel, pyrope garnet (2 wt.% Cr_2O_3), green diopside (2 wt.% Cr_2O_3), and magnesian-ilmenite. Sedimentary xenoliths include sandstone, limestone, and chert. The age of the kimberlite is unknown, but based on geological evidence, is believed to be post-Middle Ordovician (Meyer, 1976). The kimberlite intrudes Beemantown Dolomite, consists of three separate intrusives (Sweet, 1997), and has a surface area of about 700 by 1200 feet.

Bulk samples recovered from the kimberlite in the early 1980s by Cominco American Incorporated yielded no diamonds (Janse, 1994b).

Spotsylvania County

A diamond was reportedly found in the Whitehall mine in the west-central part of the county, north of State Road 608 (Sweet, 1997).

Tazewell County

In 1913, a beautiful blue-white diamond was reported from Tazewell County more than 100 miles southwest of the Mt. Horeb kimberlite in western Virginia (Holden, 1944). The gem cut from the original stone weighed 0.83 carat (Sinkankas, 1959).

Warren County

The most northerly of the known ultramafic intrusives in Virginia lies near Front Royal near the West Virginia border. This rock was initially described as a mica-peridotite dike; more recently, Southworth and others (1992) referred to this intrusive as kimberlite. The Front Royal intrusive is located about 75 miles northeast of the Staunton kimberlite. According to Sweet (1997), the intrusive occupies a surface area of about 600 by 150 feet.

West Virginia

Much of West Virginia is located within the Appalachians. One of the largest diamonds in the United States, known as the Punch Jones, was found in West Virginia. Potential diamondiferous source rocks found in the state include peridotite as well as kimberlite.

Berkeley County

A kimberlite dike discovered near Martinsburg in eastern West Virginia was initially classified as pyroxenite (Southworth and others, 1992). This kimberlite lies along a northeasterly trend with kimberlitic intrusives to the southwest in Virginia. The possibility of other kimberlitic intrusives along this trend should be considered.

Monroe County

According to the Washington Post (February 6, 1964), the Punch Jones diamond, which weighed 34.46 carats, was found in Peterstown, West Virginia, near the Virginia state line. This was the second largest verified diamond found in the United States (Hausel, 1995a, b).

The diamond was found in 1928, but was not identified until 1943 (Sinkankas, 1959). The stone was a hexoctahedron that was slightly greenish-gray and exhibited several possible impact features suggesting extensive transportation distances (Holden, 1944; Sinkankas, 1959). The source of the diamond is unknown, but the presence of kimberlite and peridotite in nearby Virginia as well as in West Virginia should be investigated, as should potential undiscovered source rocks. The diamond was placed in the U.S. National Museum along with other American diamonds.

Pendleton County

Peridotite is reported in Pendleton County near the Virginia-West Virginia state line (Blank, 1935; Southworth and others, 1992).

Great Lakes region

Parts of the Great Lakes region are underlain by the Superior Province, an Archean craton underlying much of Minnesota and eastern South and North Dakota. This archon extends north into Canada. The Superior Province is bounded on the west by the Trans-Hudson orogen, and Early to Middle Proterozoic basement (proton) to the east and south (**Figure 5**).

The Early to Middle Proterozoic basement along the eastern and southern margin of the Superior archon underlies much of the Great Lakes region. This basement is bounded by Late Proterozoic rocks of the Grenville tecton farther to the east. The Grenville tecton extends into eastern Michigan and Indiana.

Because the Great Lakes region is underlain by Archean and Early to Middle Proterozoic basement rocks, this region is considered favorable terrane for the discovery of diamondiferous kimberlite and lamproite. Several diamonds (including some fairly sizable stones) have been recovered from the Great Lakes region. Historically, these diamonds were thought to have been transported from Canada by continental glaciers during the last ice age. This assumption has recently been questioned following the discovery of several kimberlites in Michigan. It is likely that some of the diamonds found in glacial till in the Great Lakes region originated from nearby kimberlites.

Illinois

Much of Illinois is underlain by basement rocks which form part of a protonic terrane. Near the southern margin of this proton and extending into the Grenville orogen, is a swarm of lamprophyric dikes in southern Illinois near the Kentucky state line. Besides these dikes, a recent article in the Northern Miner (September 4, 1995) suggested that kimberlite had recently been found in northern Illinois (see Michigan, page 64). However, this discovery has not been verified. There are also unverified reports of diamonds being found in the state.

Hardin County

A dike swarm of more than 50 ultramafic rocks occurs in southern Illinois near Elizabethtown along the Kentucky border, and continues south into western Kentucky. These were originally described as mica-peridotite or lamprophyre and occasionally referred to as kimberlite, although their mineralogy does not support a kimberlitic classification (Meyer, 1976). Several of the intrusives were discovered during drilling in southern Illinois for Missouri Valley-type fluorite-sphalerite-galena mineralization (Koenig, 1956).

The dikes intrude Mississippian to Pennsylvanian sedimentary rocks including coal, and have a general northwesterly trend. The thermal effects on the coal suggest a maximum intrusive temperature of 600°C (Meyer, 1976). The dikes rarely exceed 25 feet in width, and are porphyritic with abundant phlogopite. Olivine, which forms much of the groundmass, is replaced by serpentine and calcite. Pyroxene is uncommon, and magnetite, ilmenite, garnet, perovskite, apatite, and chromite are minor constituents of some dikes.

Many of the dikes are associated with magnetic anomalies. Zartman and others (1967) report Permian (257 Ma) ages for some of the intrusives.

Jefferson County

A 7.75-carat stone was reportedly found 3.4 miles east of Ashley in southern Illinois. The discovery was not confirmed (Sinkankas, 1959).

McDonough County

A 1911 report indicated that 22 diamonds were found near Macomb in western Illinois. The report was not confirmed (Sinkankas, 1959).

Indiana

The western part of Indiana is underlain by a proton, which is bounded on the east by the Grenville orogen (tecton). Several diamonds were reportedly found in central Indiana by prospectors panning for gold (Blank, 1934). The source of the diamonds is unknown, but it has been assumed that the diamonds were transported by glaciers originating from Canada. However, with the recent discoveries of diamondiferous kimberlite in Michigan, the source of the diamonds should be re-evaluated.

Brown County

Several diamonds were discovered in Brown County of south-central Indiana, including a 4-carat stone (Blank, 1934). One of the Brown County diamonds examined by Holden (1944) weighed 1.69 carats and is now part of the U.S. National Museum's collection.

In 1916, a 1.48-carat, yellowish rounded dodecahedron was recovered from Lick Creek about 14 miles southeast of Martinsville. A 3.06-carat diamond was found in Salt Creek in northeastern Brown County (Sinkankas, 1959).

Miami County

In 1949, a 3.93-carat flattened and distorted octahedron was discovered near Peru in north-central Indiana (Sinkankas, 1959).

Morgan County

A 3-carat stone was found near central Indiana in 1878 in Little Indian Creek by prospectors panning for gold. Another diamond, called the Stanley diamond, was found in a branch of Gold Creek. The Stanley diamond weighed 4.88 carats and was a greenish-yellow octahedron.

Other reported stones found in the county weighed from less than 0.13 carat to 1.66 carats and consisted of dodecahedral and hexoctahedral crystals of white, yellow, bluish, and pink colors (Blank, 1934). These stones were recovered from gold pan concentrates along with some rubies, sapphires, and zircons (Sinkankas, 1959).

According to Sinkankas (1959), a 1.0-carat diamond was also found in 1908. This was followed in 1911 by a discovery of a 0.135-carat diamond at the junction of Gold and Sycamore Creeks. The same locality produced a 2.28-carat diamond in 1912. In 1913, five diamonds were washed from Gold and Highland Creeks. These included a greenish 0.2-carat diamond, a colorless 0.73-carat diamond, and a yellowish, twinned 0.69-carat diamond.

Michigan

For the most part, Michigan is underlain by a proton considered favorable for the discovery of diamondiferous kimberlite and lamproite. In recent years, Michigan has received a fair amount of diamond exploration interest, which has led to the discovery of several diamondiferous and barren kimberlites. Before this exploration program, most of the diamonds found in the Great Lakes region were considered to have been transported by glaciers from the Canadian Shield.

At least one large diamond has been found in a glacial moraine in Michigan. Morainal material from the same glacier also yielded several diamonds in Wisconsin (**Figure 51**). Until recently, all of these diamonds were assumed to have originated from Canada, but the recent discovery of diamonds in several kimberlites in the region suggests that some of these diamonds may have had a nearby source.

To date, more than 20 post-Ordovician kimberlites have been discovered in Michigan (Carlson and Floodstrand, 1994). According to the Northern Miner (September 4, 1995), Crystal Exploration identified at least 26 kimberlites in the Michigan, Wisconsin, and northern Illinois area of which eight contain subeconomic amounts of diamonds. In addition, eleven magnetic anomalies were identified that are suggestive of buried kimberlite pipes. Michigan also has some Paleozoic outliers that are completely surrounded by Proterozoic-age rocks, and currently are interpreted as cryptovolcanic structures possibly related to kimberlite. Similar Paleozoic outliers found in Wyoming and Colorado in the early 1960s lead to the discovery of kimberlite in the Colorado-Wyoming State Line district (see Wyoming craton, pg. 9).

Baraga County

Sherman Hill (section 7, T51N, R34W) in Baraga County along Lake Superior, is an outlier of deformed Paleozoic dolomite surrounded by Proterozoic sandstone. The outlier occurs as a semi-circular ridge of possible cryptovolcanic origin (Cannon and Mudrey, 1981).

Cass County

The large diamond, known as the Dowagiac diamond, was found in Michigan near the Indiana border. The diamond was a rounded hexoctahedron that weighed 10.875 carats (Sinkankas, 1959). According to Blank (1934), this stone and the Burlington and Saukville stones from Wisconsin were recovered from the Lake Michigan glacial moraine.

Houghton County

Limestone Mountain, located about 1.2 miles southwest of Sherman Hill, is another outlier of Paleozoic rock in sections 13, 14, 23, and 24, T51N, R35W. The outlier consists of dolomite surrounded by Proterozoic sandstone and forms a prominent bluff 330 feet above the surrounding topography. The structure may be cryptovolcanic (Cannon and Mudrey, 1981).

Iron County

Another outlier was recognized along the Michigan-Wisconsin border in section 27, T42N, R35W about 5 miles south of Iron River, Michigan. This outlier (the Brule River outlier as shown on **Figure 51**) is about 30 feet in diameter and surrounded by Proterozoic metavolcanic rocks (Cannon and Mudrey, 1981).

Kimberlite was initially discovered in Michigan near Crystal Falls, one mile west of Lake Ellen (SW section 27, T44N, R31W) near the Wisconsin border (**Figure 51**). The intrusive, known as the Lake Ellen kimberlite, is poorly exposed but produced a strong positive magnetic anomaly indicating it has a circular plan, is about 650 to 950 feet in diameter, and covers about 20 acres (**Figure 52**). The kimberlite was emplaced in volcanic rocks of the Hemlock Formation (Proterozoic age) and contains abundant Ordovician (?) dolomite xenoliths.

The host rock consists of diatreme facies kimberlite composed of olivine, pyroxene, mica, pyrope, and magnesian ilmenite in a fine-grained serpentine matrix (**Figure 53**) (Cannon and Mudrey, 1981). Small diamonds have been recovered from the Lake Ellen intrusive. Dow Chemical constructed a diamond extraction plant in Crystal Falls, which is part of a joint venture with Crystal Exploration and Ashton Mining. Another kimberlite, known as the Michgamme kimberlite, lies a short distance northwest of the Lake Ellen kimberlite along the Michgamme Reservoir shoreline (Carlson and Floodstrand, 1994).

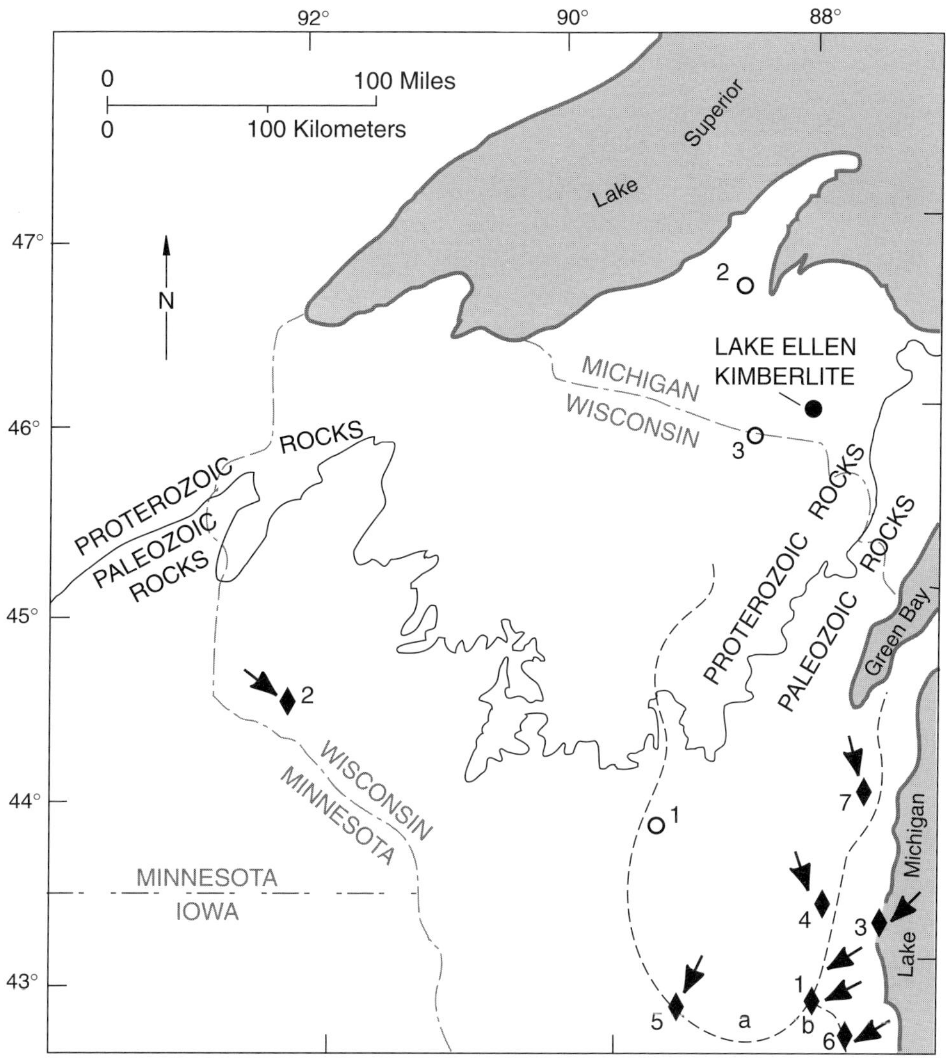

EXPLANATION

○ Possibble cryptovolcanic structures
1. Glover Bluff
2. Structures near Pelkie (includes Limestone Mountain and Sherman Hill)
3. Brule River outlier

---- Extent of ice lobes
a. Green Bay lobe
b. Lake Michigan lobe

♦ Drift diamond localities
1. Eagle, Waukesha County
2. Plum Creek, Pierce County
3. Saukville, Ozaukee County
4. Kohlsville, Washington County
5. Oregon, Dane County
6. Burlington, Racine County
7. Collins, Manitowoc County

➘ Direction of ice movement near diamond localities

Figure 51. Location map of diamond occurrences, possible cryptovolcanic structures, and the Lake Ellen kimberlite, Michigan (after Cannon and Mudrey, 1981).

Ohio

Much of Ohio is underlain by basement rocks of the Grenville orogen. Only two diamonds were reportedly found in Ohio. One was discovered near Cleveland in the northern part of the state (Mather, 1941), and the second diamond was found in Clermont County in the southern part of the state. The latter diamond weighed 6 carats and was found in 1897 at Milford near Cincinnati (Blank, 1934). The diamond was found in the "Kettle" moraine.

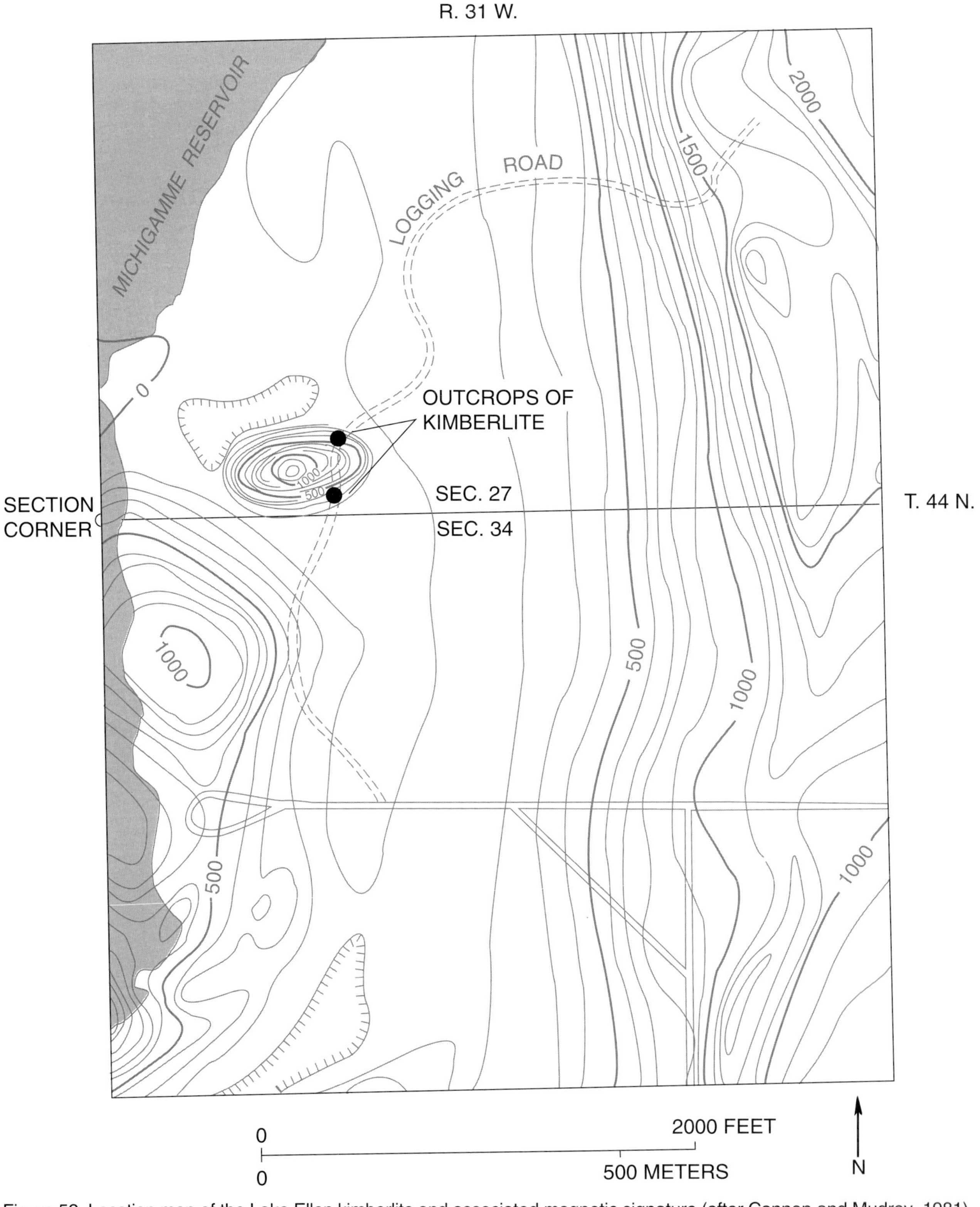

Figure 52. Location map of the Lake Ellen kimberlite and associated magnetic signature (after Cannon and Mudrey, 1981).

Wisconsin

Northwestern Wisconsin is underlain by basement rocks of the Superior Province, while Proterozoic rocks underlie much of the remainder of the state. Since 1876, approximately 25 diamonds have been found in seven regions in southern and central Wisconsin (**Figure 51**). All these diamonds were found in Pleistocene glacial deposits or in Holocene river gravel. The presence of cryptovolcanic structures and the apparent discovery of kimberlite in

Figure 53. Sample of diatreme facies kimberlite from Lake Ellen, Michigan.

Wisconsin, as well as a kimberlite swarm in nearby Michigan (Northern Miner, September 4, 1995), raises the possibility that some of the Wisconsin diamonds were derived from nearby kimberlites.

Dane County

In 1893, a white diamond was discovered near Oregon, Wisconsin, on the Johnstown moraine in southern Wisconsin. The diamond was found on Judson Devine's place 2.5 miles southwest of Oregon (Blank, 1934). The stone weighed 4.0 carats (Cannon and Mudrey, 1981). Sinkankas (1959) reported this diamond as a 3.87-carat gray-green, distorted and rounded octahedron that was recovered from the "Kettle" moraine.

Langlade County

In 1984, Al Falster, an amateur mineralogist, found 5 diamonds near Antigo in northeastern Wisconsin (Bendheim, 1984).

Manitowoc County

Several uncut diamonds were found in a coffee pot belonging to an old hermit named Peter Zagloba, who lived in Collins in eastern Wisconsin. Following his death, the diamonds were discovered by neighbors. The diamonds possibly originated from gravel in the Collins area (Cannon and Mudrey, 1981).

Marquette County

Cannon and Mudrey (1981) suggested that the Glover Bluff structure (SW section 3, T17N, R8E) in Marquette County of south-central Wisconsin (**Figure 51**), could possibly be related to a diatreme. The structure is roughly circular, 1600 feet in diameter, and encloses Cambrian and Ordovician strata that have been fragmented and down-dropped 200 feet.

The Glover Bluff structure lies near a prominent east-west trending positive magnetic anomaly and within an area surrounded by a positive Bouguer gravity anomaly. The anomalies are interpreted to be the expression of a mafic intrusion at a depth of about one mile below the surface.

Ozaukee County

A diamond discovered near the Milwaukee River in 1881, known as the Saukville diamond, was a beautiful white, flattened trisoctahedron of 6.57 carats (Cannon and Mudrey, 1981). The diamond was found three miles north of Saukville and one mile from the Lake Michigan shoreline (Blank, 1934).

Pierce County

Between 1880 and 1887, approximately 10 small diamonds were recovered from the gold-bearing gravels along Plum Creek in the Rock Elm Township, western Wisconsin (Blank, 1934; Cannon and Mudrey, 1981). The diamonds weighed less than a carat each. In 1889, three other diamonds, greenish-gray in color, were recovered from the west branch of Plum Creek. These crystals were hexoctahedral and weighed 0.8, 0.44, and 0.03 carats, respectively (Sinkankas, 1959). These latter diamonds may have been part of a fraud (Cannon and Mudrey, 1981).

Exploration in this region in the early 1980s identified a potential cryptovolcanic structure expressed through geophysical signatures. Stream sediment samples collected from the area recovered coarse gold nuggets and scheelite. Apparently no kimberlite or lamproite was found.

Racine County

In 1903, a 2.11-carat stone was discovered in glacial drift near Burlington in southeastern Wisconsin. The diamond was a twinned, flattened, tetrahedron of faintly greenish-gray color (Blank, 1934).

Washington County

The Terresa diamond was discovered at Kohlsville in 1883 in southeastern Wisconsin a few miles west of the Saukville diamond in adjacent Ozaukee County. The Terresa diamond weighed 21.25 carats (reported as 21.5 carats by Cannon and Mudrey, 1981) and was found near the Green Lake moraine. The diamond was nearly spherical with a flaw or cleavage that separated the diamond into a colorless crystal on one side, and a cream-yellow crystal on the opposite side (Cannon and Mudrey, 1981). The diamond was cut in 1918 and produced a total of 9.27 carats of finished stones, the largest being 1.48 carats (Sinkankas, 1959).

Waukesha County

According to Kunz (1885), a 15-carat diamond (reported as 16.25 carats by Cannon and Mudrey, 1980) was dug from a 60-foot-deep well in glacial drift at Eagle, southeastern Wisconsin, in 1883. The diamond was slightly off color (warm yellow) (Kunz, 1885). According to Sinkankas (1959), the Eagle diamond was a dodecahedron that weighed 15.37 carats, and was found in the Kettle moraine in 1876. Kunz (1885) reported that the diamond was recovered from hard, yellowish ground in the well. Two other diamonds were also found in this area that weighed less than 0.5 carat each.

Continental Interior region

The Continental Interior region, including Kansas, Missouri, Nebraska, and South Dakota, is underlain primarily by stable Middle Proterozoic (proton) and Late Proterozoic basement rocks. This region includes several known kimberlites, lamproites, and some related ultramafic rocks. Diamonds, however, are uncommon, but the abundance of kimberlite and lamproite in eastern Kansas suggests that similar intrusions may be discovered in this region in the future.

Kansas

Eleven kimberlites and several lamproites intrude sedimentary strata in eastern Kansas. Other than two small diamonds that were possibly recovered from one of the Kansas kimberlites (William Mansker, written communication, 1994), no other diamonds have been reported in Kansas. The kimberlite group is situated in Riley County near the southeastern flank of the Mid-continent geophysical anomaly and is underlain by Proterozoic mafic rocks. This geophysical anomaly has been interpreted as the result of an aborted paleorift.

Lamproites in the state are located in Woodson County south of the kimberlites and include olivine lamproite with a rock chemistry similar to the diamondiferous lamproites found in Arkansas and Western Australia. The sedimentary rocks of the stable platform in Kansas overlie stable Proterozoic-age basement rocks.

Riley County

To date, eleven kimberlite intrusives have been recognized in Riley County of northeastern Kansas (**Figure 54**). They intrude limestones and shales of the Chase Group formations and were emplaced at about 120 Ma and possibly as early as 95 Ma (Brookins, 1970a, c). Since emplacement, the intrusives have not been subjected to a large amount of erosion, inasmuch as crater facies, diatreme facies, and hypabyssal facies kimberlite are present. The presence of crater facies kimberlite is uncommon and has only been reported at a few localities in the world; it is reported in three of the Riley County kimberlites (Baldwin Creek, Fancy Creek, and Winkler).

The known kimberlite intrusives in Kansas include in alphabetical order, Bala, Baldwin Creek, Fancy Creek, Leonardville, Lone Tree A and B, Randolph 1 and 2, Stockdale, Swede Creek, and Winkler (Mansker and others, 1987). There are also reports of pyrope garnet recovered from water well

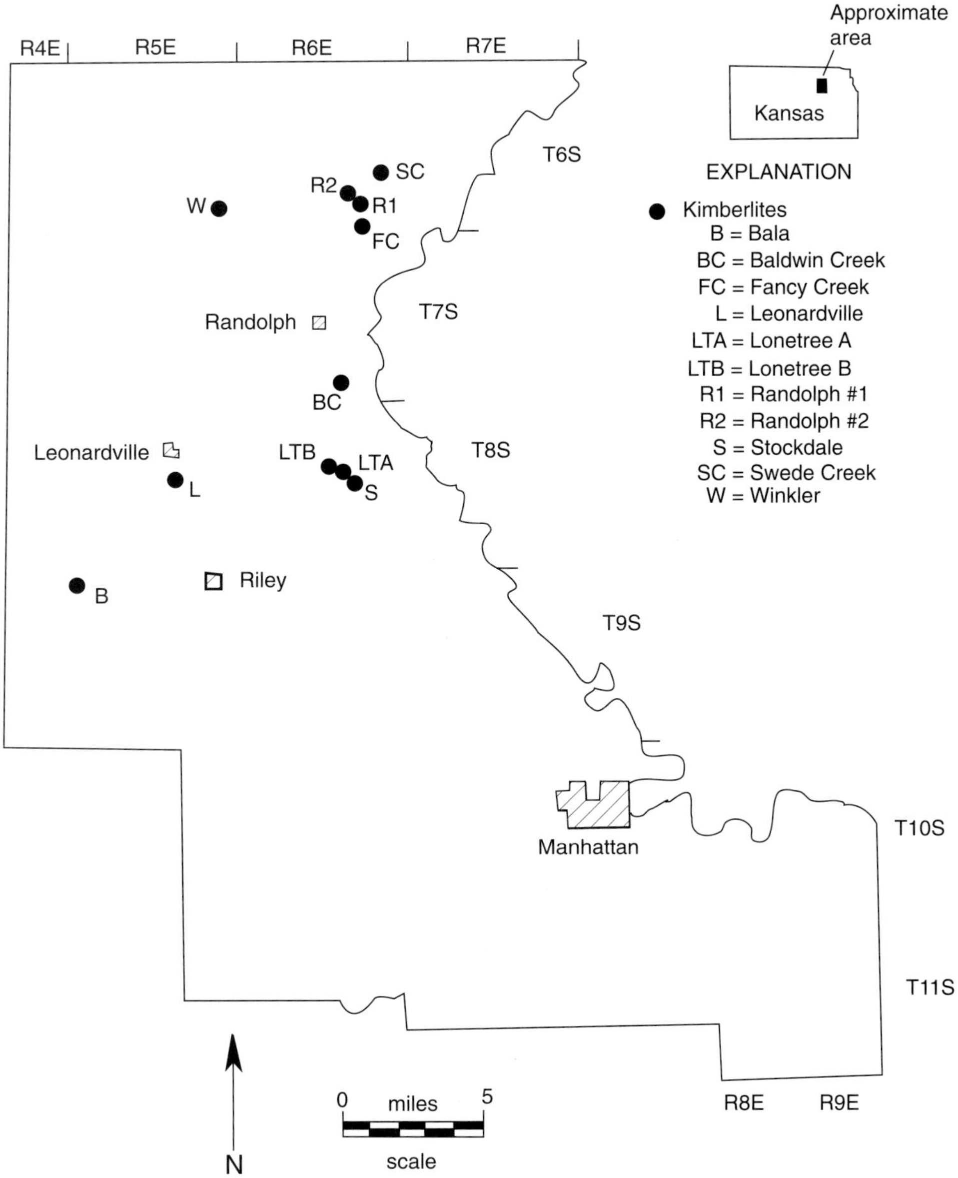

Figure 54. Map of Riley County, Kansas, showing locations of known kimberlites (after Mansker and others, 1987).

cuttings about 0.5 mile west of the Winkler crater, which suggests the presence of an undiscovered kimberlite pipe or sill in that area (Brookins, 1970a).

Magnetic data indicates that several of the Kansas kimberlites possess mushroom-like caps caused by lateral spreading near final sites of emplacement. Six of the intrusives yielded strong positive magnetic anomalies (Brookins, 1970c; Mansker and others, 1987). For instance, ground surveys detected anomalies in the range of 500 to 6320 gammas, and aeromagnetic anomalies in the range of 30 to 200 gammas (Mansker and others, 1987). The known intrusives are expressed as low-lying knolls, with a few intrusives marked by distinct topographic depressions.

Bala kimberlite. The Bala kimberlite is located 5 miles west of Riley in NW NW section 6, T9S, R5E, and forms a rounded knoll approximately 15 feet high and 200 feet in diameter. The kimberlite is hypabyssal facies and contains abundant ilmenite, magnetite, and serpentinized olivine phenocrysts as well as country rock xenoliths (Brookins, 1970c).

Baldwin Creek intrusive. This intrusive is located about 2 miles south of the Randolph community in E/2 section 34, T7S, R6E. The Baldwin Creek intrusive is not exposed at the surface, and is buried under the Baldwin Creek floodplain. The intrusive was identified by its magnetic signature (750 gammas) which suggests the presence of an elongate, 300-foot-wide intrusive (Mansker and others, 1987).

Fancy Creek kimberlite. This kimberlite is located in W/2 W/2 section 2, T7S, R6E. The Fancy Creek kimberlite was discovered following an airborne magnetic survey by Cominco American Incorporated in the early 1980s. The intrusive produced a 50-gamma anomaly which was accentuated by a ground anomaly of 1060 gammas. The intrusive forms a topographic low with a poorly exposed outcrop along the edge of a cow pond that was constructed in the middle of a depression in the 900-foot by 750-foot crater (**Figure 55**). Kimberlitic indicator minerals are common in the weathered outcrop next to the cow pond as well as in the material used to build the dam (Mansker and others, 1987) (**Figure 56**).

The kimberlite consists of highly brecciated and comminuted tuffaceous crater facies kimberlite and diatreme facies kimberlite. Two diamonds may have been recovered from the intrusive by Cominco American Incorporated in the early 1980s (Jim Marin, personal communication, 1994; William Mansker, written communication, 1994), although the diamonds may have been contaminates from samples processed from the Colorado-Wyoming State Line district.

Leonardville. This kimberlite is located about one mile south of the Leonardville community in NW section 22, T8S, R5E. The kimberlite consists of diatreme facies and produced a ground magnetic high of about 2000 gammas (Mansker and others, 1987). The kimberlite forms two mounds along a gently sloping hill and consists of highly porphyritic and brecciated diatreme facies kimberlite. Accidental and cognate xenoliths are abundant (Brookins, 1970c). The intrusive is possibly 100 feet across.

Lone Tree A and B. The Lone Tree A and B kimberlites, located in N/2 section 23, T8S, R6E, lie a short distance northwest of the Stockdale kimberlite and about 5 to 6 miles northeast of the town of Riley. The three intrusives lie along a northwesterly trend, suggesting that all three intrusives share a common structural control at depth. The two Lone Tree kimberlites consist of diatreme facies, produce strong ground magnetic anomalies of more than 2000 gammas, and are relatively small. For instance, Lone Tree A is about 350 feet in diameter and Lone Tree B is only about 35 feet in diameter.

A nearby pond west of Lone Tree A and south of Lone Tree B contains decomposed kimberlite as well as some kimberlitic indicator minerals (Mansker and others, 1987).

Randolph 1 and 2. Randolph 1 is located in SE section 27, and Randolph 2 in the NW section 35, T6S, R6E. These intrusives lie 4 to 4.5 miles north of the Randolph community and form relative topographic highs. Randolph 1 forms a circular knoll about 200 feet in diameter and produced a 6300-gamma ground magnetic high. Randolph 2 also forms a topographic high and is about 50 feet in diameter. The intrusives consist of hypabyssal facies kimberlite (Brookins, 1970c; Mansker and others, 1987).

Stockdale kimberlite. This kimberlite is located in NE and SE section 23, T8S, R6E. The Stockdale kimberlite is poorly exposed but consists of diatreme facies

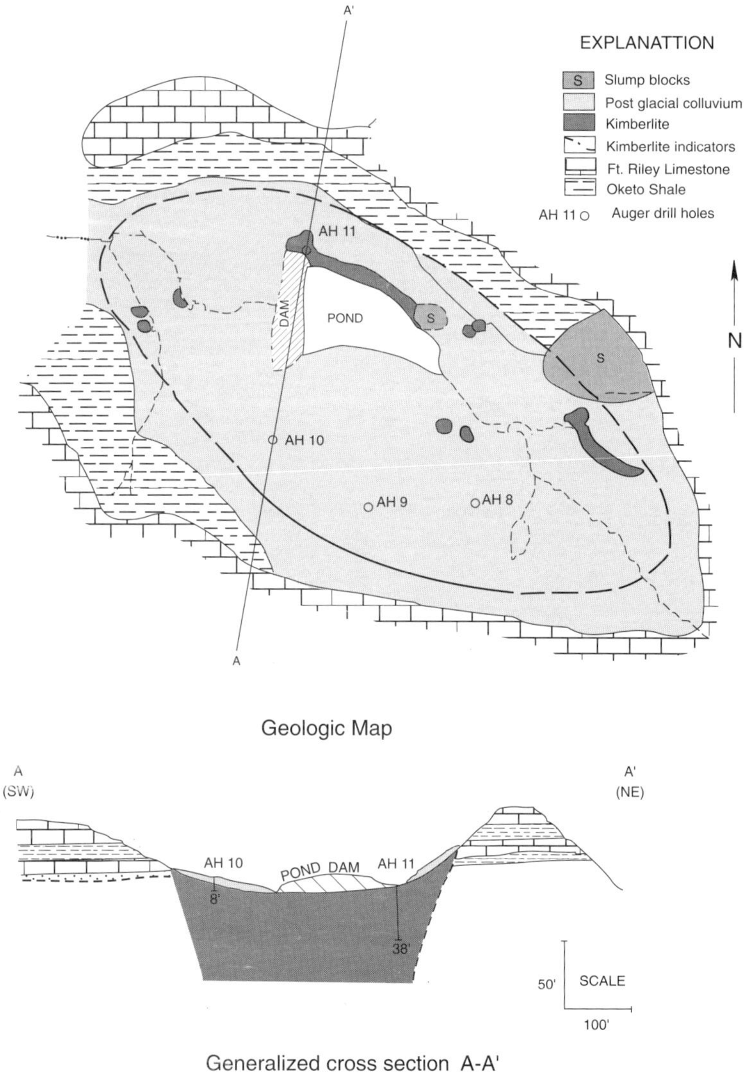

Figure 55. Generalized geologic map and cross section of the Fancy Creek kimberlite, Riley County, Kansas (after Mansker and others, 1987).

a.

b.

Figure 56. (a) View of topographic depression over Fancy Creek kimberlite, Riley County, Kansas, and (b) two diamonds possibly recovered from the intrusion (photographs courtesy of Junior Nelsen).

kimberlite intruding Lower Permian sedimentary rocks (**Figure 57**). This intrusive was described by Rosa and Brookins (1967) as consisting of olivine, pyroxene, ilmenite, pyrope, chloritized phlogopite, magnetite, and calcite in a highly serpentinized and carbonated matrix. A variety of cognate xenoliths (some of which are deep-seated) includes eclogites, lherzolites, pyroxenites, magnetites, and granulites (Brookins and Woods, 1970a). Nodules from the Stockdale are similar to those recovered from the Monastery Mine, South Africa, and according to Brookins and Woods (1970b) establishes a minimum depth of genesis of 70 miles. Brookins (1970a) suggested that the depth of formation of the Riley County kimberlites was approximately 70 to 90 miles.

Some of the nodular ilmenite-pyroxene intergrowths from the Stockdale kimberlite are assumed to have formed at even greater depths, possibly as much as 180 miles, although other researchers have placed a minimum depth of formation at 70 miles (Brookins, 1970a). Diamond is assumed to be stable at depths of 96 miles at 1500° C, although some researchers suggest stability at depths of 70 to 90 miles and temperatures of 1000°C to 1400°C.

Figure 57. View of exposed Stockdale kimberlite, Riley County, Kansas.

Accidental xenoliths of Paleozoic sedimentary country rock are common in the Stockdale, Leonardville, Bala, and Winkler kimberlites. These xenoliths consist of shales, limestones, and cherts. Lower crustal xenoliths include granite, adamellite, quartzite, diorite, gabbro, and schist (Mansker and others, 1987).

Swede Creek. This kimberlite is located about 1.5 miles northeast of the Randolph 2 kimberlite and about 6 miles north of the Randolph community in W/2 section 24, T6S, R6E. This was a buried intrusive of hypabyssal facies kimberlite that was exposed in a trench. A ground magnetic high of about 6000 gammas was detected over the intrusive (Mansker and others, 1987).

Winkler kimberlite. Located in N/2 section 36, T6S, R5E, the Winkler kimberlite forms a 950-foot diameter circular crater that was listed as a possible impact structure in 1964. This topographic anomaly is similar to some of the Lonetree topographic anomalies recently identified in the southern Green River Basin of Wyoming (see Uinta County, Wyoming, page 38). A ground magnetic anomaly of 550 gammas was detected over the structure, and was later drilled. According to Brookins (1970b), the 100-foot drill hole encountered yellow ground at a depth of 17 feet, blue ground at a depth of 21 feet, and diatreme and crater facies kimberlite at 85 feet.

Although kimberlite does not appear to be exposed at the surface, sample concentrates from a poorly defined drainage running through the center of the structure yielded pyrope garnet, chromian diopside, and picroilmenite. This suggests that the kimberlite was exposed to erosion.

Woodson County

Lamproite sills and vents occur in the southeastern part of the state. The Hills Pond and Rose Dome lamproites occur 80 miles east of Wichita, and a few miles south of Yates Center in Woodson County (37.667°N, 95.750°W). These lamproites are dated at 88 to 91 Ma, and intrude Precambrian granite and flat-lying Pennsylvanian shales and limestones that are part of a broadly domed Late Paleozoic platform overlying stable Proterozoic-age (1.3 to 2.0 Ga) cratonic rocks. The Hills Pond intrusive is exposed in scattered outcrops (**Figure 58**) and was being mined by Microlite Corporation for fertilizer additive in 1994. The Rose Dome lamproite is not exposed at the surface, and was discovered by drilling. Several other unnamed lamproites are reported nearby (Cullers and others, 1985; Markezich, 1985).

The Hills Pond lamproite is a coarse-grained, porphyritic, olivine-phlogopite-richterite-diopside madupitic lamproite; the Rose Dome lamproite (Silver City dome) is an altered and carbonated olivine-phlogopite madupitic lamproite. These lamproites are olivine lamproites and consist of Ti-phlogopite, serpentinized olivine, K-richterite, and Ti-diopside, in a fine-grained groundmass of serpentine, perovoskite, apatite, and chrome spinel. Samples obtained by drilling the Hills Pond intrusive are intensely altered to vermiculite (Bergman, 1987).

There are seven other known significant occurrences of olivine lamproite in the world. All, except the Woodson County lamproites, are diamondiferous (Rock and others, 1992). The geochemistry of the Woodson County lamproites is identical to the diamondiferous lamproites at Ellendale and Argyle,

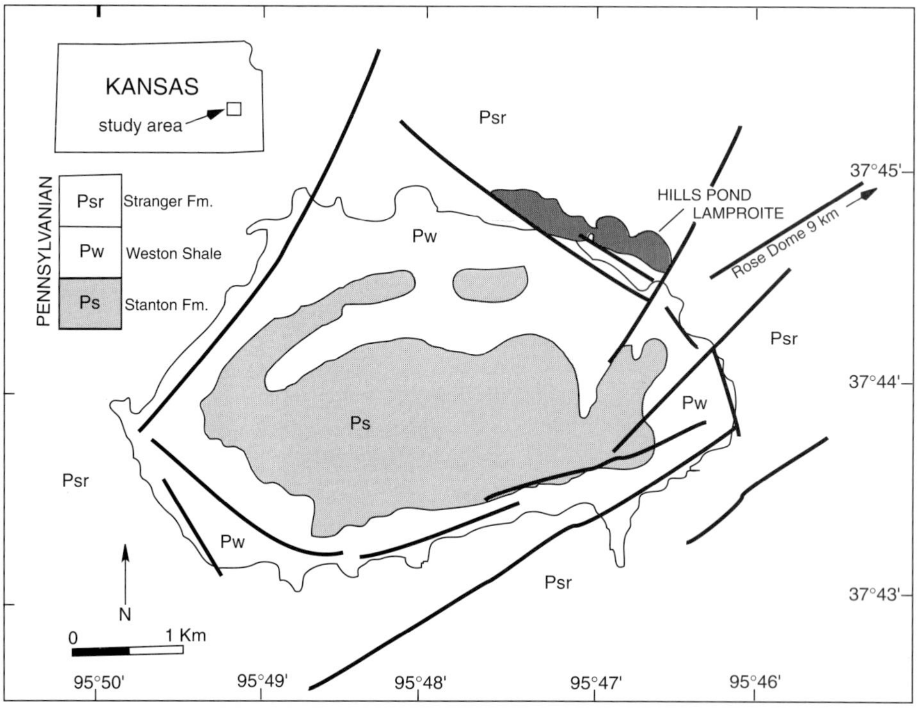

Figure 58. Generalized sketch map of the Silver City dome, Arkansas, showing location of the Hills Pond lamproite (modified from Wojcki and Knapp, 1990).

Western Australia, and Prairie Creek, Arkansas (Bergman, 1987).

Bulk sampling of the Hills Pond and the Rose Dome intrusives by Cominco American Incorporated in the early 1980s yielded no diamonds (Coopersmith and Mitchell, 1989). It is suggested that Hills Pond and Rose Dome lamproites are unfavorable for diamond preservation, as they are sills which tended to cool slowly on emplacement. Such conditions would favor diamond resorption.

Missouri

At least 100 small diatremes and associated dikes occur in a region of southeastern Missouri around Ste. Genevieve and St. Francois Counties south of St. Louis. Early studies indicated that one of the intrusives was peridotite, although Ball and Singewald (1930) classified the same intrusion near Avon as alnöite based on the presence of abundant melilite in the groundmass. This intrusion was hot enough to metamorphose the surrounding Cambrian dolomite for a distance of about 20 feet from the contact. The presence of Devonian fossils in limestone xenoliths in some of the intrusions was used to date the time of emplacement as post-Devonian.

Mansker (1973) examined the petrology and mineralogy of one pipe near Avon, Missouri, and reported finding olivine, clinopyroxene, phlogopite, and chromian spinel in the pipe. Notable was the absence of pyrope garnet and picroilmenite, although an earlier study by Kidwell (1947) on 50 of the intrusions identified garnet as an accessory mineral. Melilite was found in only one intrusion, contrary to Ball and Singewald (1930). Mansker (1973) also identified xonolite, which due to its low thermal stability would suggest a low temperature of emplacement.

The intrusions yielded radiometric ages of 377 to 399 Ma (Early to Middle Devonian) (Erlich and others, 1989). According to Brookins and others (1979), some of the diatremes near Avon have kimberlitic affinity. Missouri is underlain by Proterozoic age basement rock considered to be part of a stable proton.

Nebraska

The Elk Creek carbonatite in Nebraska is reported to contain kimberlitic material (Brookins and others, 1979).

South Dakota

Western South and North Dakota are underlain by basement rocks of the Trans-Hudson orogen. These include Archean rocks that were affected by a thermal event which largely reset isotopic compositions to Proterozoic ages (Goldich and others, 1966). The eastern parts of both states are underlain by Archean rocks of the Superior Province.

There are no known kimberlites, lamproites, or diamonds in South Dakota, although there was an early report of diamonds near Yankton near the Missouri River in southeastern South Dakota. Mather's (1941) statement: "Doubtful reports of diamonds from Yankton... " remains to be checked.

Gulf Coast region

The Gulf Coast region includes a large region along the Gulf Coast and includes Arkansas, Florida, Louisiana, Mississippi, Oklahoma, and Texas, and parts of Alabama and Georgia. Much of Arkansas and parts of Texas are underlain by Early to Middle Proterozoic basement rocks considered to be part of a proton, and thus are favorable for diamondiferous lamproite and kimberlite.

In particular, olivine lamproite of the Prairie Creek intrusive in Arkansas, located along the edge of the Ouachita Mountains, has been relatively productive for diamonds. The lamproite was the site of North America's first diamond mine.

Arkansas

Diamonds were first discovered in 1906 near the mouth of Prairie Creek, 2.5 miles southeast of Murfreesboro in Pike County, southwestern Arkansas. This area has been one of the two most productive areas for diamond in the United States and has produced more than 90,000 diamonds, including the largest verified diamond found in the United States at 40.42 carats.

Most of the diamonds were recovered from an intrusive which has been incorporated into the Crater of the Diamonds State Park. The host rock was initially referred to as peridotite and later as kimberlite, although recent studies show that the whole rock chemistry and mineralogy is typical of olivine lamproite. In the same area, at least five other lamproites have been recognized.

Peridotite has been recovered from drill core in Scott, Cleveland, and Ashley Counties, to the north, east, and southeast, respectively, of Murfreesburo (Bolivar, 1984, 1988), and kimberlite is reported in Scott County (Salpas and others, 1986). No diamonds have been reported from these intrusives. Elsewhere in the state, alnöites and monchiquites have been reported to occur.

Howard County

Gravel lenses in Mine Creek, four miles north of Nashville (section 2, T9S, R27W) in southwestern Arkansas, contain pebbles of quartzite, novaculite, grit, syenite, tinguaite, as well as fourchite (an olivine-free lamprophyre). The gravel was tested for diamonds with negative results (Miser, 1914).

Pike County

Diamonds were discovered in olivine lamproite at Prairie Creek near the town of Murfreesburo in 1906 (**Figure 59**). Shortly following the discovery, the property was prospected and mined on a small scale for diamonds, although the operations proved to be uneconomical (**Figure 60**). The property is now included in the Crater of the Diamonds State Park, but has recently been reevaluated for economic mineralization (Mike Howard, personal communication, 1996).

Prairie Creek lamproite. The olivine lamproite at Prairie Creek located in sections 21 and 28, T8S, R25W, near Murfreesburo, has been one of the two most productive regions in the United States for diamonds. Estimated production is from 90,000 to 100,000 diamonds. These diamonds included 30% gemstones and 70% industrial stones; no attempt has ever been made to recover microdiamonds (Sinkankas, 1959). The grade of the pipe has been estimated at 11 carats/100 tonnes with an average diamond weight of 0.26 carat.

Some of the larger diamonds recovered from this pipe include the Uncle Sam (40.42 carats), the Star of Murfreesboro (34.25 carats), the Amarillo Starlight (16.37 carats), and the Star of Arkansas (15.24 carats) (**Table 1**). Most of the diamonds are white, yellow, or brown, and the most common form is a distorted hexoctahedron with rounded faces (Bolivar, 1984; Kidwell, 1990).

The Prairie Creek olivine lamproite intrudes rocks of the Cretaceous Trinity Formation, and lies marginal to the Ouachita orogen within the Gulf Coast region, which is characterized by Cretaceous age sedimentary rocks that dip gently to the south (Meyer and others, 1977). The age of the pipe, based on geological evidence, is Late Cretaceous, which agrees with a 106-Ma K-Ar age date on phlogopite (Gogineni and others, 1978). This region of Arkansas, including parts of Texas and Oklahoma, appears to have been subjected to considerable alkalic igneous activity during the Cretaceous.

The diamond pipe covers approximately 73 acres and contains breccia, tuff, and hypabyssal olivine lamproite (Miser and Ross, 1922a; Bolivar, 1984) (**Figure 61**). Nearly all the diamonds have been recovered from the breccia facies of the lamproite.

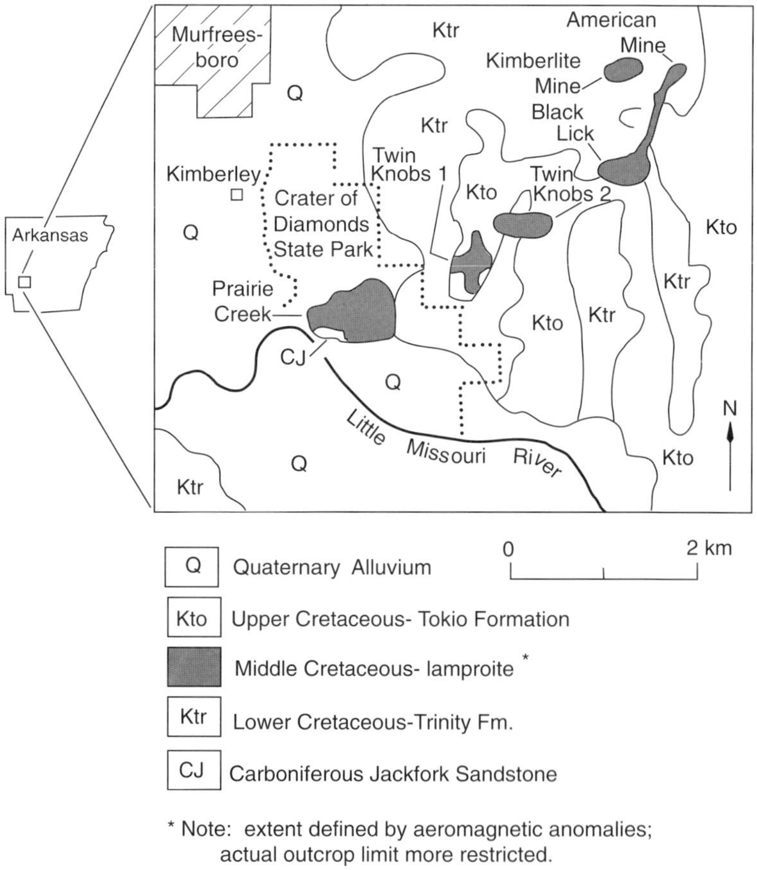

Figure 59. Geologic map of the Prairie Creek, Arkansas lamproite field (modified from Krol, 1988).

a.

b.

Figure 60. Remains of early mining activity at Murfreesburo, Arkansas, showing old headframe (a) and concentrating plant (b).

Figure 61. Tuffaceous olivine lamproite breccia, Prairie Creek, Arkansas.

The rock contains diopside, olivine, chrome spinel, phlogopite, potassium richterite, perovskite, pyrite, pyrrhotite-pendlandite, and uncommon pyrope and almandine. Much of the olivine is altered to serpentine. Country rock xenoliths include shale, limestone, and sandstone. No upper mantle xenoliths have been found in the intrusive. Gogineni and others (1978) reported that pyrope analyses showed compositions equivalent to G9 calcic-chrome pyropes which were probably derived from the disaggregation of garnet lherzolite from the upper mantle. Fipke and others (1995) reported that one sub-calcic G10 pyrope garnet was found in their samples, and that chromite analyses were unfavorable, as none of the chromites recovered in their samples yielded geochemistry similar to diamond inclusion geochemistry.

Other lamproites. Other lamproites that lie approximately 2 miles northeast of the Prairie Creek intrusive have been identified in the area (**Figure 59**). These include the Kimberlite (NE section 14, T8S, R25W), American (SW NW section 14, T8S, R25W), Black Lick (NW NW section 23, T8S, R25W), Twin Knobs 1 (W/2 SW section 22, T8S, R25W), and Twin Knobs 2 (NE SW section 22, T8S, R25W) intrusives (Krol, 1988; Mike Howard, written communication, 1996). Other ultramafic rocks of lamproitic or lamprophyric affinity have been reported a few miles east of Prairie Creek and about 3 miles south of Corinth.

Both the Kimberlite and American lamproites have also yielded some diamonds, although these two intrusives are apparently lower grade than the Prairie Creek pipe (Miser, 1914; Miser and Ross, 1922a). The Kimberlite deposit is a hypabyssal facies porphyritic olivine lamproite, and the American deposit is exposed at the surface as "yellow ground" with a small outcrop of olivine lamproite at its center. Miser (1914) reported that several diamonds were recovered from a washing operation on the lamproite.

The Black Lick intrusive is a porphyritic olivine lamproite exposed at the surface as "black ground." No diamonds are known to have been found in this lamproite.

Pope County

An alnöite, known as the Dare Mine Knob, is reported in section 11, T11N, R20W (Mike Howard, personal communication, 1996). The rock is a fine-grained gray to dark greenish-gray alnöite.

Scott County

A kimberlitic intrusive, known as the Blue Ball kimberlite, was discovered in 1909 about 24 miles southwest of Danville in sections 3 and 4, T3N, R26W, in the bank of Freedom Creek. This sill was originally identified as mica-peridotite by Miser and Ross (1922b) and recently was reclassified as kimberlite by Salpas and others (1986), even though it lacks the typical mantle megacrysts and nodules found in most kimberlites. The sill reportedly averages about 5.5 feet thick, has been traced over a distance of 1000 feet along a N82°E trend, and is conformable to the steeply dipping (70°N) country rock shales and sandstones.

Peridotite was also found about 1/4 mile to the southwest in a 10-foot-deep pit. At this locality, the peridotite is about 1 foot wide and was believed by Miser and Ross (1922b) to represent a separate sill based on its trend and location.

Both sills were described as consisting of aggregates of serpentine and phlogopite phenocrysts in a matrix of serpentinized olivine and phlogopite with accessory calcite and magnetite. Miser and Ross (1922b) recognized that the sills were similar to the diamond-bearing intrusive at Murfreesburo; however, prospecting between 1909 and 1923 yielded no diamonds.

White County

A perfect, virtually flawless, hexoctahedral diamond was found in 1925 on the Pellie Howell farm at Searcy in eastern Arkansas. The Howell diamond weighed 27.31 carats and was sold to Tiffany and Company of New York (Sinkankas, 1959).

Oklahoma

Oklahoma, like Arkansas, is underlain by Proterozoic rocks which form part of a proton. Some pyrope garnet and chromian diopside anomalies associated with Mississippian sedimentary rocks were apparently identified in the Broken Bow region of southeastern Oklahoma several years ago. However, the source of the indicator minerals was apparently not found.

Texas

The basement rock of northern Texas forms part of a proton which extends into Oklahoma and New Mexico. The central part of Texas is apparently underlain by a tecton (Janse, 1994b) which extends south. South of the tecton, the coastal plain sediments continue to the Gulf Coast. Only the northern part of Texas would be considered potentially favorable for diamond deposits.

A few diamonds have been found in Texas; however, the source of the diamonds is unknown. One sampling program completed in Foard County, northern Texas, near where a diamond was recovered, reportedly identified some pyrope garnet trains to the west associated with some Tertiary sedimentary rocks.

Foard County

A diamond was found in 1911 in section 64, Block 44 in the northern part of the state near the Oklahoma border (Sinkankas, 1959). No details were given.

Montgomery County

Several stones were reported in 1911 in Montgomery County in southeastern Texas north of Houston. The stones reportedly weighed between 2.5 and 3.5 carats (Sinkankas, 1959).

Walker County

A 1.7-carat diamond was reported from Huntsville in southeastern Texas. This locality lies north of the Montgomery County diamonds (Blank, 1935).

Pacific Coast, western United States region

The Pacific Coast region of the United States consists of an intensely deformed and accreted terrane. Regions similar to the Pacific Coast have in the past been considered poor exploration targets for diamondiferous kimberlite and lamproite. However, the recent discoveries of several unconventional diamond deposits in similar terranes around the world suggest that this region could host some unusual diamond deposits (Hausel, 1996c). So far, diamonds have been reported from several locations in this region.

Alaska

Diamonds in Alaska include three stones found in a gold placer on Crooked Creek in the Circle mining district northeast of Fairbanks. The Circle district lies near the fragmented northern margin of the North American craton.

The diamonds were found between 1982 and 1986, and weighed 0.3, 0.83, and 1.4 carats, respectively. No kimberlitic indicator minerals were found in the area which may suggest that the diamonds originated from a lamproite or related intrusive, or that the stones may have originated from a distal source. A distal source is supported by the presence of percussion marks and fractures in the diamonds, suggesting that they had a complex alluvial history (Forbes and others, 1987).

The area in which the diamonds were found is a Tertiary basin. The basin fill is derived from Late Proterozoic through Late Paleozoic sedimentary and metamorphic rocks from the Crazy Mountains to the north, and Paleozoic to Precambrian metamorphic and Late Cretaceous granitic plutonic rocks from the Yukon-Tanana region to the south. Some alkaline igneous rocks are also reported to the south, although no kimberlites or lamproites have been identified (Forbes and others, 1987).

Two diamond inclusions were also found in a native platinum nugget at Goodnews Bay in the southwestern corner of the state. Another microdiamond was reportedly recovered from a core sample of bottom sediments in the Bay (Forbes and others, 1987). The source of the diamonds is unknown.

California

California has been the third most productive region in the United States for diamonds after the Colorado-Wyoming State Line district and Murfreesburo, Arkansas. But because this region does not fit any classical exploration model for diamondiferous kimberlite, it has been historically neglected by geologists.

Diamonds were discovered in 1849 near Placerville in El Dorado County, central California. Three years later, diamonds were discovered in the Cherokee hydraulic gold placer mine 8.5 miles north of Oroville in Butte County, California. Diamonds have also been reported in several other placer gold deposits in Amador, Nevada, Plumas, and Trinity Counties. In total, more than 600 diamonds (both gem and industrial stones) have been reported from 15 counties in California (Hill, 1972).

Most California diamonds were found north of the 36°N latitude, and nearly all the diamonds have been found in the Sierra Nevada and in the Klamath Mountains downstream from serpentinized ophiolite complexes and melanges. Less than one percent of the diamonds have been found in beach sands.

The close association of the placer diamonds with serpentinized ophiolite suggests possible preservation of diamonds in an obducted Beni Bousera-type mantle slab. Furthermore, the presence of pyrope garnet and chromian diopside in some stream gravels in northern California, and the discovery of chromian diopside-bearing pyroxenite and peridotite in northern and central California, provide evidence of some unusual rocks present at the surface. Possibly, the source of some of the diamonds in California and Oregon is from tectonically emplaced, mantle-derived peridotite, and/or mafic and ultramafic diatremes that sampled a subducted diamond-bearing oceanic slab at depth (Hausel, 1994b, 1996c).

Amador County

Five localities south of Placerville in central California have produced diamonds. At three of the localities, diamonds were produced from ancient river channels. More than 60 diamonds (the largest weighed about 1.5 carats) were recovered from Jackass Gulch at the town of Volcano from an old alluvial channel about 38 feet below a bed of volcanic ash.

At Fiddletown, diamonds were recovered from gravels underlying volcanic ash from Loafer Hill (the largest was approximately 1.33 carats). Indian Gulch, east of Amador City is also cited as a diamond locality, although few facts are available. According to Kunz (1885), four diamonds were recovered from gray cemented gravel underlying a layer of lava flows or ash.

The pale-colored Evans diamond, weighing about 1 carat, was found on the surface of gravel at Rancheria (about 4 miles northwest of Volcano) in 1883. The diamond was recovered in the SW NW section 18, T7N, R12E. In 1934, The Echols diamond, a stone weighing about 2.5 carats was found near Plymouth, about 19 miles south of Placerville.

Diamondiferous lamproite may have been found in this region. According to the Northern Miner (January 29, 1996), Diadem Resources drilled into a cluster of dikes including a 6000-foot-long, 600-foot-wide dike after following a train of indicator minerals upstream from a historic diamond placer at Leek Springs. The drill cuttings from 120 feet of the lamproite yielded 235 diamond fragments (Northern Miner, May 20, 1996).

Butte County

The most productive district for diamonds in California was the hydraulic gold placer mining area immediately north of Oroville in the Round Mountain area of Butte County (**Figure 62**). Between 1853 and 1918, about 400 diamonds and 600,000 ounces of gold were recovered from the historical mining operations situated near the Feather River.

The district is dominated by Round Mountain which forms a prominent flat-topped mountain north of Oroville. Round Mountain is capped by the propylitically altered Miocene-age Lovejoy Basalt. The basalt overlies a Tertiary (?) age, gold-bearing, diamondiferous conglomerate, which is the reported source of the majority of the diamonds recovered from the Oroville area.

Diamonds were initially washed from the conglomerate along the north end of Round Mountain at the Cherokee placer. In 1866, diamonds were also found in gravels along the west bank of the Feather River about one mile north of Oroville. Diamonds were also recovered from conglomerates at Thompson Flat, Yankee Hill, and Morris Gulch (Hill, 1972).

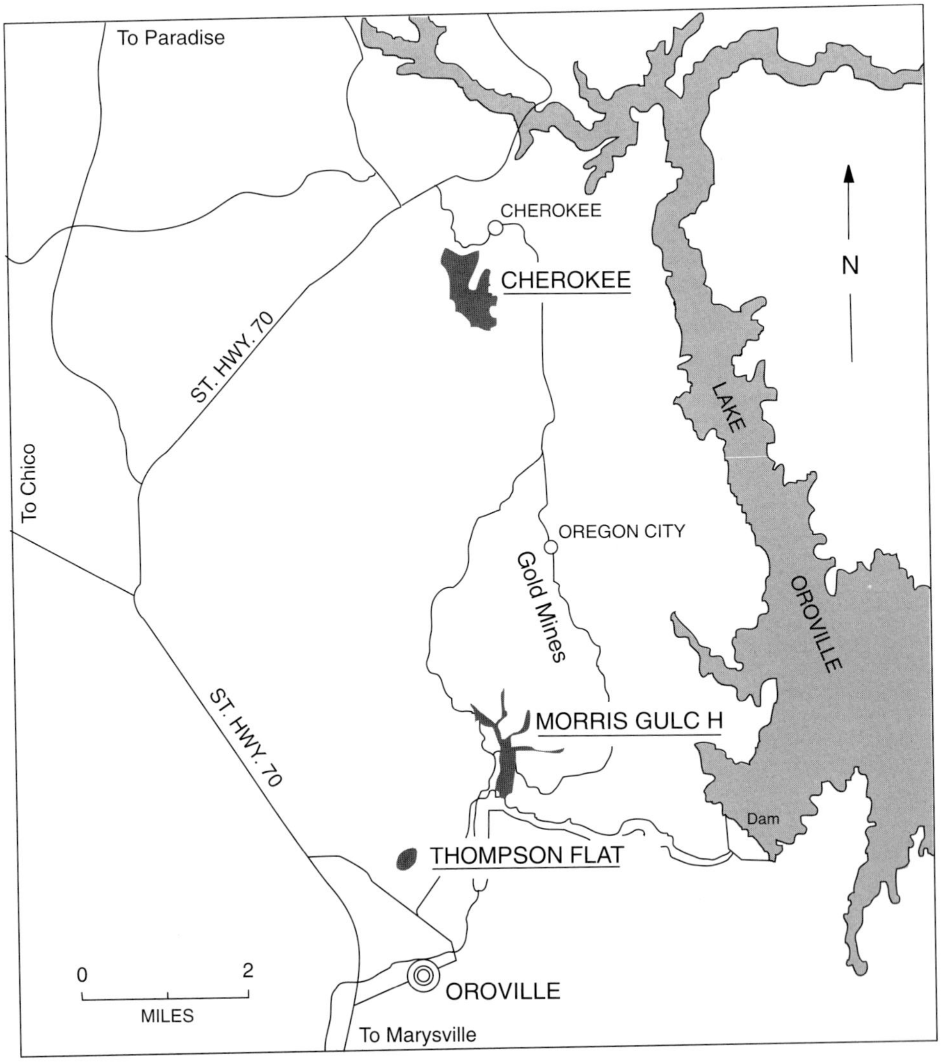

Figure 62. Location map of the Cherokee mine north of Oroville, Butte County, California (from Heylum, 1985).

In 1885, Kunz reported that 15 diamonds from the Cherokee flats were derived from a 3-foot-thick layer which was part of a larger 26-foot-thick unit of gravel. Kunz (1885) indicated that a few diamonds from this layer were of possible gem quality.

In 1906, the U.S. Diamond Company sank a 180-foot-deep shaft in what was claimed to be kimberlite. The shaft reportedly intersected a rich pocket of diamonds, although it is questionable whether any diamonds were recovered from the deposit (Rosenhouse, 1975). Presently, kimberlite is not known in the area, but serpentinized amphibolite schist has been described (Heylum, 1985).

The largest recorded diamond found in the Oroville area was found in 1868. The stone, known as the Moore diamond, weighed 2.25 carats. However, local folklore claims that the largest diamond from the area weighed 6 carats and was found outside the entrance of the Spring Valley mine at Thompson Flat (Rosenhouse, 1975). The color of the recovered stones included pure white, rose, and yellow colored diamonds (Hill, 1972).

At Thompson Flat, two miles north of Oroville, diamonds were found in placer diggings in a Tertiary river channel and a few stones were recovered from nearby modern drainages (Hill, 1972). Diamonds from the Oroville area have been surprisingly good quality being relatively free of flaws and possessing good brilliance. However, many have a yellowish tinge which detracts from their value (Heylum, 1985).

In 1931, another diamond weighing 0.5 carat was found in the Cherokee mine by a prospector. Another resident reported finding a diamond that weighed 2 carats and 27 points in the same year (Blank, 1934). Kunz (1885) reported that the diamondiferous gravels at Cherokee flats also included platinum, almandine garnet, epidote, gold, iridosmine, limonite, magnetite, pyrite, quartz, rutile, topaz, and zircon.

Del Norte County

Kunz (1885) reported that diamonds may be found in all the northern counties in California drained by the Trinity River; in the vicinity of Coos Bay, Oregon; and on the banks of the Smith River of Del Norte County, California. Sinkankas (1959) reiterated that microdiamonds had been found on the Smith River.

El Dorado County

Records of diamonds from El Dorado County are very complete due to the diligent efforts of a Judge Carpendar who kept notes on the recovered diamonds. Diamonds from the Placerville region included stones from Live Oak Mine on Reservoir Hill, Smiths Flat, Spanish Ravine, White Rock Ravine, Webber Hill, Wisconsin Flat mine, Texas Hill, Newtown, Fairplay, and Prospect Flat. At least 50 diamonds were recovered from this region. Diamonds described by Judge Carpendar were white, canary yellow, greenish, and blue, weighing from 0.1 carat to 1.82 carats. Of the 50 diamonds described by Carpendar, at least 13 stones weighed more than 0.95 carat (Hill, 1972). Some of the diamonds may have been gem quality, in that Kunz (1885) indicated some of the stones had been cut and mounted in rings.

A diamond from Forest Hill was described as weighing 1.5 carats and was of good color. The diamond was found in a tunnel driven into the auriferous gravels (Kunz, 1885). The diamondiferous gravel was described as capped by a bed of lava ranging from 50 to 450 feet thick.

Fresno County

Small diamonds have been reported a few miles north of Coalinga (Blank, 1934).

Humboldt County

Diamonds were reported in the Trinity River in Humboldt County (Sinkankas, 1959).

Imperial County

A diamond was reported near the Mexican border. The discovery was never authenticated (Sinkankas, 1959).

Nevada County

One of the largest diamonds found in California was a 7.25-carat stone recovered from French Corral (Heylum, 1985). A second stone from this area weighed 1.33 carats. A stone described by Kunz (1885) weighed 1.25 carats, was remarkably free of flaws, and slightly yellowish in color.

Plumas County

Diamonds ranging from microscopic to about 2 carats in weight have been reported at four localities in Plumas County. These include Gopher Hill, Spanish Creek, Sawpit Flat, and Nelson Point. At the latter two localities, the diamonds were recovered from Tertiary fluvial gravels.

San Diego County

Diamonds have been reported from Ramona. Three diamonds were recovered from a sluice box on Hatfield Creek in the flats of the Little Three Mine. The stones were reported to be 1/8th-inch in diameter (Sinkankas, 1959).

Siskiyou County

Diamonds were reported in the placer gravels of the Hamburg bar (Blank, 1934).

Trinity County

Five large diamonds were recovered from a tributary of the South Fork of the Trinity River known as Hayfork Creek (**Figure 63**). One of these diamonds, named the Doubledipity, weighed 32.99 carats, and was found in 1987. This was the fourth largest authenticated diamond found in the United States (**Table 1**). The Doubledipity is a yellowish-brown diamond and lacked the typical adamantine luster commonly associated with diamond. It is opaque except on the edges where it is translucent and does not fluoresce. The stone is an aggregate of seven interpenetrating cubes in random crystallographic orientation (Kopf and others, 1990).

In addition to the Doubledipity, at least four other large diamonds were recovered from Hayfork Creek in the Klamath Mountains of northern California. These included the Enigma (17.83 carats), the Serendipity (14.33 carats), a poorly documented diamond discovered in the 1860s that was about half an inch in diameter and estimated to weigh between 10 to 15 carats, and the Jeopardy diamond (3.9 carats).

These large stones all exhibit an overgrowth of a multitude of tiny crystal faces of diamonds which coat the underlying crystal or crystal aggregate. The recognized forms are both cubes and octahedrons. The morphology of the crystals suggests that they were subjected to disequilibrium conditions within the diamond stability field. The diamonds, being industrial, are coated with encrustations and form aggregates which suggest minimal transport distance from the source rock (Kopf and others, 1990).

The Enigma diamond is a grayish-brown industrial stone that is semi-translucent to opaque as a result of microscopic black specs scattered throughout the diamond. The stone is triangular in shape and bounded by two cleavage surfaces, indicating the stone was originally larger (Kopf and others, 1990).

Along with these large diamonds, countless numbers of small diamonds have been reported in the black sands of the Trinity River. Sinkankas (1959) reported that microdiamonds were found in the black sands of the Trinity River near its junction with the Klamath River.

The placers in this region also carry significant gold, coarse-grained platinum-group metals, and chromite. Recently, pyrope garnet and chromian diopside (heavy mineral indicators of mantle material) were also reported from the Trinity River area (Kopf and others, 1990). Both of these minerals tend to disaggregate during stream transportation over short distances, and the combination of these minerals suggests a nearby ultramafic or ultrabasic source terrane(s) for the diamonds.

Tulare County

A single crystal was found in 1895 from Alpine Creek. The diamond was never authenticated (Sinkankas, 1959).

Oregon

Diamonds were found in the black sands near Coos Bay, Oregon (Kunz, 1885), and in placer gold gravels of southwestern Oregon (Anonymous, 1981). The number of recovered diamonds was estimated to be about 100 stones, and most were yellowish in color. The stones include some good quality diamonds up to 3 carats in weight, with larger stones of inferior quality (Blank, 1934). One unusual report indicated that a 3.15-carat gemstone had been found in the craw of a chicken in Oregon (Birdsall, 1986). Microscopic diamonds were reportedly found in beach sands at several localities along the coast of Oregon (Anonymous, 1961). Essentially all Oregon diamonds found to date have been in placers downstream from alpine and ophiolitic peridotites.

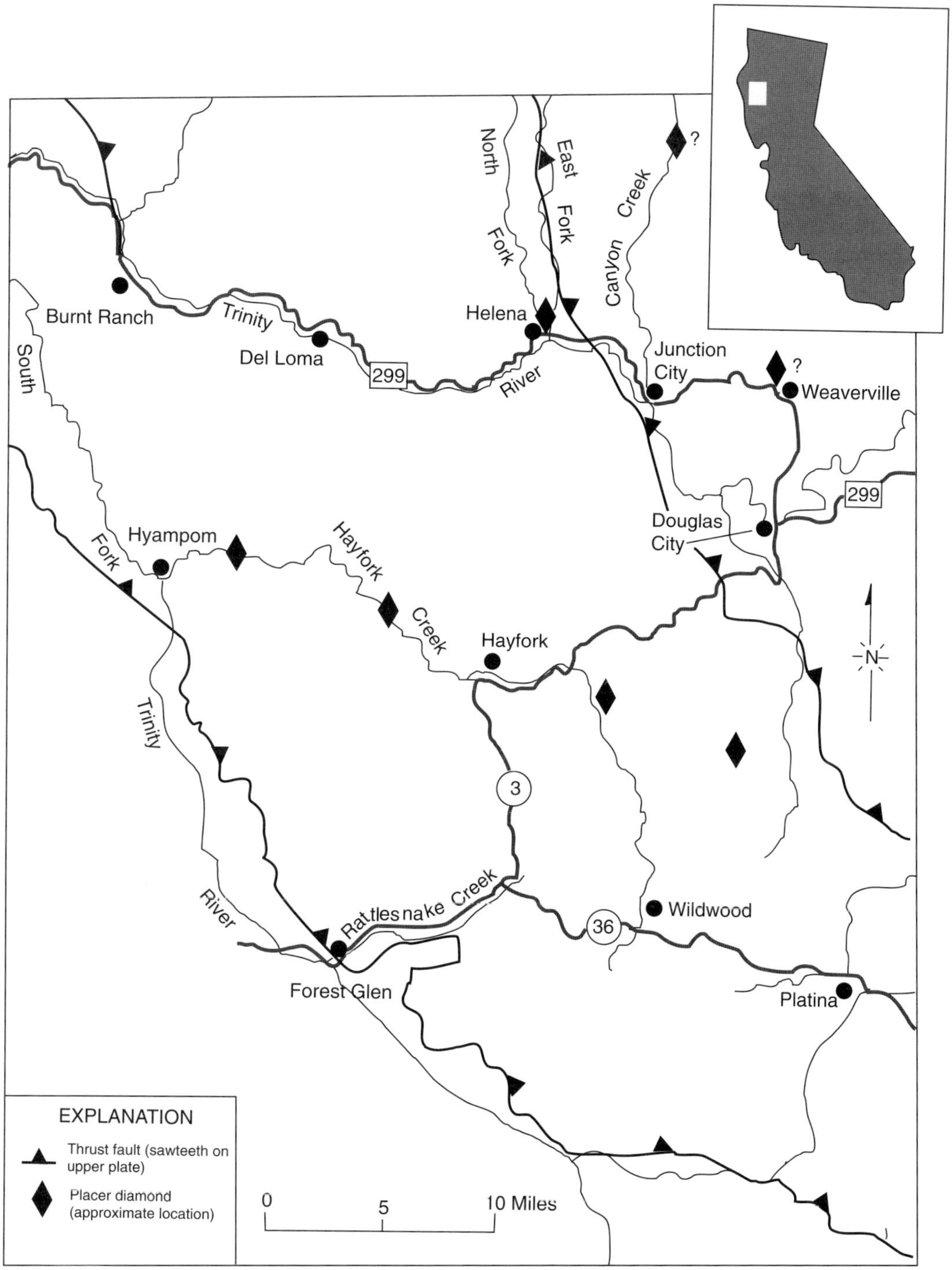

Figure 63. Location map of diamond localities along the Hayfork Creek and Trinity River, northern California (after Kopf and others, 1990).

Curry County

A diamond found near Wedderburn near the mouth of the Rogue River reportedly weighed 0.60 carat. The diamond was a flattened hexoctahedron with minor carbon inclusions (Anonymous, 1981). The diamond was incorporated into the U.S. National Museum's collection (Holden, 1944).

Josephine County

A diamond recovered from a gold placer in the county weighed nearly 3 carats (Anonymous, 1961). Two additional poor-quality yellow diamonds were reported from Josephine Creek (Birdsall, 1986).

Malheur County

A 2.5-carat diamond was found in 1870 in a gold placer. The diamond was dark in color and described as bort (Anonymous, 1961; Birdsall, 1986).

Washington

Microscopic diamonds have been reported in black sand deposits along the coast in the northwestern part of the state (Blank, 1934). The geology of the coast of Washington is similar to Oregon and northern California; thus, one might anticipate similar unconventional sources for diamond in Washington and Oregon.

In 1932, a diamond of a little more than 4 carats was found in a Skamania County gold placer. The stone was a pale yellow octahedron (Blank, 1934). No other reports of diamonds are known from Washington.

Conclusions

Presently, the only commercial diamond deposits in the United States are located in the Colorado-Wyoming State Line district in the western U.S. Two kimberlites in this district, the Kelsey Lake KL1 and KL2 diatremes, were placed into production by Redaurum Red Lakes Ltd. in 1996. This mine, located in Colorado and adjacent to the Wyoming state line, is located within a field of nearly 40 known diamondiferous kimberlites, and the possibility of other commercial deposits in this district is considered good. In the past, the only other commercial diamond deposit in the U.S. was a diamondiferous lamproite at Murfreesburo, Arkansas. This deposit was mined near the turn of the 19th century, and was evaluated in 1996 as a potential source for gemstones.

Considering the geology, the presence of favorable rock types, and reported diamond occurrences, it is apparent that six areas or regions in the United States stand out as potentially favorable. These are: (1) the Appalachian Mountains region, (2) the Great Lakes region, (3) Kansas, in the Continental Interior region, (4) Arkansas, in the Gulf Coast region, (5) the Wyoming craton in the Rocky Mountain-Great Basin region, and (6) parts of the Pacific Coast region. The area considered most favorable is the Wyoming craton.

Numerous diamonds and some kimberlites have been found in the Appalachian Mountains. The emplacement of kimberlite dikes and diatremes in this region are controlled by intersections of reactivated faults with cross-lineaments (Parrish, 1984). Although exploration in this region would be hindered by heavy vegetation, the potential for discovery of additional kimberlites is considered to be high. Exploration should include mapping lineaments, stream-sediment sampling surveys, airborne geophysical surveys, as well as following up on the historical reports of indicator minerals and diamond.

Because the Great Lakes region is underlain by cratonic basement rocks, several kimberlites and some cryptovolcanic structures have been identified in the area. In addition to the several large diamonds reported, some diamondiferous kimberlites have been found in the region. This region may represent one of the better exploration targets for diamonds in the United States.

Kansas is also considered a favorable exploration target. Although reports of diamonds are uncommon, possibly because of the lack of historical placer gold mining in the state, the presence of several kimberlites and lamproites in eastern Kansas suggests there is a possibility for the discovery of additional pipes. Like the Appalachian Mountains, Kansas is covered by extensive grasslands and thick stands of trees. Stream-sediment sampling in conjunction with airborne geophysical surveys would provide the best exploration program in this region.

Thousands of diamonds have been recovered from a small group of lamproites in Murfreesburo, Arkansas. This region is heavily vegetated, making exploration difficult, but there is still potential for discovery of additional diamondiferous lamproite in this region. The use of airborne magnetic surveys in the Arkansas region may lead to the discovery of additional lamproites.

Using only the number of anomalies identified, the Colorado-Wyoming kimberlite province stands out as a potentially important diamond province. To date, many of the diamondiferous kimberlites that have been identified in this region have been considered economically marginal. However, several large gemstones have recently been recovered from this region, and ore grades in some of the kimberlites are favorable. The region also hosts numerous kimberlitic anomalies that need to be examined in greater detail. The abundance of kimberlite, lamproite, and kimberlitic indicator mineral anomalies in this region overlying an archon and proton should make the Colorado-Wyoming kimberlite province one of the highest priority regions for diamond exploration in North America.

Part of the Pacific Coast region (northern California and southern Oregon) provides an entirely different opportunity for the exploration geologist. This region is virgin ground for diamond exploration, yet it may offer potential for the discovery of an unconventional diamond source rock. More than 600 diamonds that range from microdiamonds to the fifth largest diamond ever found in the U.S., have been recovered from this region. Many of these diamonds were recovered under very unfavorable conditions (i.e., hydraulic mining) not conducive to diamond recovery. Undoubtedly, hundreds, and possibly thousands of diamonds reported to the tailings during the historic hydraulic mining operations in this region.

Much of the northern California coast near the diamond discoveries is tectonically unstable, and parts of this region are underlain by a Phanerozoic basement, conditions that have in the past been considered antipathetic to diamonds. However, the discovery of diamonds in ophiolite and in alpine peridotite in the former Soviet Union and in Tibet, the discovery of diamonds in alkali basalt in New South Wales, and the recent discovery of abundant graphitized diamonds in ophiolite in Morocco, all emphasize that this region along the Pacific Coast may be a potentially important region for future diamond exploration.

In summary, the numerous reported diamonds, kimberlites, and lamproites in the United States suggest that this country could become a significant diamond producer in the future if more emphasis were placed on diamond exploration.

References cited

Allen, J.E., and Bulk, R., 1954, Mineral resources of Fort Defiance and Tohatchi Quadrangles, Arizona and New Mexico: New Mexico Institute of Mines and Mineral Resources Bulletin 36, p. 100-118.

Anderson, O.L., and Perkins, P.C., 1975, A plate tectonics model involving non-laminar asthenospheric flow to account for irregular patterns of magmatism in the southwestern United States: Institute of Geophysics and Planetary Physics, University of California, Los Angeles, Publication 1241, p. 113-122.

Anonymous, 1961, Diamonds in Oregon: The Mineralogist, July-August, p. 15.

Anonymous, 1981, Oregon diamond find has been verified: Oregon Geology, v. 43, no. 3, p. 1.

Anonymous, 1990a, Uncut 14-carat diamond found by Montana jogger: Associated Press, September 10, 1990.

Anonymous, 1990b, Diamond find raises hope in Craig: Associated Press, Missoulian, November 19, 1990.

Anonymous, 1994, Young diamond reveals gem's priceless quality: Geotimes, v. 39, no. 10, p. 5.

Atkinson, W.J., 1989, Diamond exploration philosphy, practice, and promises: a review, *in* Ross, J., and others, editors, Kimberlites and related rocks, Volume 2, Their mantle/crust setting, diamonds and diamond exploration, Proceedings of the fourth international kimberlite conference, Perth, Australia, 1986: Geological Society of Australia Special Publication 14, p. 1075-1107.

Ball, S.H., and Singewald, J.T., 1930, Discussion of "An alnöite pipe, its contact phenomena, and ore deposition near Avon, Missouri": Journal of Geology, v. 38, p. 456-459.

Barton, M., and van Bergen, M.J., 1981, Green clinopyroxenes and associated phases in a potassium-rich lava from the Leucite Hills, Wyoming: Contributions to Mineralogy and Petrology, v. 77, p. 101-114.

Bendheim, R., 1984, Diamonds in them thar glacial deposits???: Wisconsin Geological and Natural History Survey, Survey View, v. 5, no. 2, p. 3.

Bergman, S.G., 1987, Lamproites and other potassium-rich igneous rocks: a review of their occurrence, mineralogy and geochemistry, *in* Fitton, J.G., and Upton, B.G.J., editors, Alkaline igneous rocks: Geological Society Special Publication no. 30, p. 103-190.

Birdsall, L., 1986, Although extremely rare, diamonds occur in several Oregon localities: The Oregonian, October 30, p. E2.

Blank, E.W., 1934, Diamond finds in the United States: Rocks and Minerals, v. 9, p. 147-182.

Blank, E.W., 1935, Diamond finds in the United States: Rocks and Minerals, v. 10, no. 2, p. 23-26.

Bolivar, S.L., 1982, Kimberlite of Elliott County, Kentucky: Kentucky Geological Survey Thesis Series 2, 33 p.

Bolivar, S.L., 1984, An overview of the Prairie Creek intrusion, Arkansas: Society of Mining Engineers of AIME Transactions, v. 280, p. 1988-1993.

Bolivar, S.L., 1988, An overview of the Prairie Creek intrusion, Arkansas: Society of Mining Engineers of AIME Preprint 84-364, 12 p.

Bolivar, S.L., and Brookins, D.B., 1979, Geophysical and Rb-Sr study of the Prairie Creek, Arkansas, kimberlite, *in* Boyd, F.R., and Meyer, H.O.A., editors, Kimberlites, diatremes, and diamonds: Their geology, petrology, and geochemistry, Volume 1, Proceedings of the second international kimberlite conference: American Geophysical Union, Washington, D.C., p. 289-299.

Braddock, W.A., and Cole, J.C., 1990, Geologic map of Rocky Mountain National Park and vicinity, Colorado: U.S. Geological Survey Miscellaneous Investigations Series Map I-1973, scale 1:50,000.

Braddock, W.A., Cole, J.C., and Eggler, D.H., 1989, Geologic map of the Diamond Peak Quadrangle, Larimer County, Colorado, and Albany County, Wyoming: U.S. Geological Survey Geologic Quadrangle Map GQ-1614, scale 1:24,000.

Brookins, D.G., 1970a, Factors governing the emplacement of Riley County, Kansas kimberlites: Geological Survey of Kansas Bulletin 199, part 4, 17 p.

Brookins, D.G., 1970b, Kimberlite at Winkler Crater, Kansas: Geological Society of America Bulletin, v. 81, p. 241-246.

Brookins, D.G., 1970c, The kimberlites of Riley County, Kansas: Geological Survey of Kansas Bulletin 200, 32 p.

Brookins, D.G., Della Valle, R.S., and Bolivar, S.L., 1979, Significance of uranium abundance in United States kimberlites, *in* Boyd, F.R., and Meyer, H.O.A., editors, Kimberlites, diatremes, and diamonds: Their geology, petrology, and geochemistry, Volume 1, Proceedings of the second international kimberlite conference: American Geophysical Union, Washington, D.C., p. 280-287.

Brookins, D.G., and Woods, M.J., 1970a, High pressure mineral reactions in a pyroxenite granulite nodule from the Stockdale kimberlite, Riley County, Kansas: Geological Survey of Kansas Bulletin 199, part 3, 6 p.

Brookins, D.G., and Woods, M.J., 1970b, Rb-Sr geochronologic investigation of basic and ultrabasic xenoliths from the Stockdale kimberlite, Riley County, Kansas: Geological Survey of Kansas Bulletin 199, part 2, 12 p.

Brown, W.R., 1977, Geologic map of the Williard Quadrangle, northeastern Kentucky: U.S. Geological Survey Geologic Quadrangle Map GQ-1387, scale 1:24,000.

Bruton, E., 1979, Diamonds: Chilton Book Company, Radnor, Pennsylvania, 532 p.

Cannon, W.F., and Mudrey, M.G., Jr., 1981, The potential for diamond-bearing kimberlite in northern Michigan and Wisconsin: U.S. Geological Survey Circular 842, 15 p.

Carlson, J.A., 1983, Exploration for kimberlite and geophysical delineation of diatremes, W. State Line district, Colorado/Wyoming: M.S. thesis, Colorado State University, Fort Collins, 180 p.

Carlson, J.A., and Marsh, S.W., 1989, Discovery of the George Creek kimberlite dikes, Colorado, *in* Ross, J., and others, editors, Kimberlites and related rocks, Volume 2, Their mantle/crust setting, diamonds and diamond exploration, Proceedings of the fourth international kimberlite conference, Perth, Australia, 1986: Geological Society of Australia Special Publication 14, p. 1169-1178.

Carlson, S.M., and Floodstrand, W., 1994, Michigan kimberlites and diamond exploration techniques: Institute on Lake Superior Geology, Proceedings, 40th Annual Meeting, May 11-14, Michigan Technological Institute, 15 p.

Carmichael, I.S.E., 1967, The mineralogy and petrology of the volcanic rocks from the Leucite Hills, Wyoming: Contributions to Mineralogy and Petrology, v. 15, p. 24-66.

Carter, B.A., 1982, Geology of the Eocene volcanic sequence, Mt. Baldy-Union Peak area, central Garnet Range, Montana: M.S. thesis, University of Montana, Missoula, 43 p.

Cook, R.B., 1978, Minerals of Georgia: their properties and occurrences: Georgia Department of Natural Resources, Atlanta, Georgia, p. 19-20.

Coopersmith, H.G., 1991, Geology and exploration of the Kelsey Lake diamondiferous kimberlites, Colorado: Society of Mining Engineers Preprint 91-174, 17 p.

Coopersmith, H.G., 1993, Diamondiferous kimberlite at Kelsey Lake, southern Wyoming Archean province, *in* Dunne, K. and Grant, B. editors, Mid-Continent diamonds: Geological Association of Canada Mineral Deposits Division Symposium Volume, May 17-18, Edmonton, Alberta, p. 85-87.

Coopersmith, H.G., 1997, The Kelsey Lake diamond project: the first American diamond mine: Paper presented at 100th National Western Mining Conference, Denver, Colorado, 12 p.

Coopersmith, H.G., and Mitchell, R.G., 1989, Geology and exploration of the Rose lamproite, south-east Kansas: *in* Ross, J., and others, editors, Kimberlites and related rocks, Volume 2, Their mantle/crust setting, diamonds and diamond exploration, Proceedings of the fourth international kimberlite conference, Perth, Australia, 1986: Geological Society of Australia Special Publication 14, p. 1179-1191.

Coopersmith, H.G., and Schulze, D.J., 1996, Development and geology of the Kelsey Lake diamond mine, Colorado, *in* Thompson, T.B., editor, Diamonds to gold–I. State Line kimberlite district, Colorado, II. Cresson mine, Cripple Creek district, Colorado: Society of Economic Geologists Guidebook Series, p. 5-19.

Cross, W., 1897, Igneous rocks of the Leucite Hills and Pilot Butte, Wyoming: American Journal of Science, v. 4, no. 20, p. 115-141.

Cullers, R.L., Ramakrishnan, S., Berendsen, P., and Griffin, T., 1985, Geochemistry and petrogenesis of lamproites, Late Cretaceous age, Woodson County, Kansas, U.S.A.: Geochimica et Cosmochimica Acta, v. 49, p. 1383-1402.

Deines, P., 1968, The carbon and oxygen isotopic composition of carbonate from a mica peridotite dike near Dixonville, Pennsylvania: Geochimica et Cosmochimica Acta, v. 32, p. 613-625.

Doden, A.G., 1996, Ultramafic lamprophyres from Porcupine Dome, east-central Montana, and their potential for diamond (abstract): 32nd Annual Forum on the Geology of Industrial Minerals, Laramie, Wyoming, Abstracts with Program, p. 14-16.

Doden, A.G., 1997, Ultramafic lamprophyres from Porcupine Dome, east-central Montana, and their potential for diamond, *in* Jones, R.W., and Harris, R.E., editors, Proceedings of the 32nd annual forum on the geology of industrial minerals: Wyoming State Geological Survey Public Information Circular 38, p. 241-256.

Doden, A.G., and Gold, D.P., 1993, Kimberlite xenocrysts from the Porcupine Dome diatremes, east-central Montana: evidence for multiple sources of garnet and evaluation of diamond potential: Geological Society of America Abstracts with Programs, v. 25, no. 5, p. A98.

Dummett, H.T., Fipke, C.E., and Blusson, S.L., 1987, Diamond exploration geochemistry in the North American cordillera, *in* Elliott, I.L., and Smee, B.W., editors, GEOEXPO/86, Exploration in the North American Cordillera: Association of Exploration Geochemists, Vancouver, British Columbia, p. 168-176.

Eggler, D.H., 1967, Structure and petrology of the Virginia Dale ring-dike complex, Colorado-Wyoming Front Range: Ph.D. dissertation, University of Colorado, Boulder, 153 p.

Eggler, D.H., and Furlong, K.P., 1991, Petrochemical and geophysical evidence for old mantle lithosphere beneath Montana, *in* Baker, D.W., and Berg, R.B., editors, Guidebook of the central Montana alkalic province–Geology, ore deposits and origins: Montana Bureau of Mines and Geology Special Publication 100, p. 87-91.

Eggler, D.H., Meen, J.K., Welt, F., Dudas, F.O., Furlong, K.P., McCallum, M.E., and Carlson, R.W., 1988, Tectonomagmatism of the Wyoming Province: Colorado School of Mines Quarterly, p. 25-40.

Ellsworth, P.C., 1996, Submittal summary for the Bearmouth lamproite diamond prospect: Consultants report, Missoula, Montana, July 27, 7 p.

Engineering and Mining Journal, 1994, Exploration roundup: v. 195, no. 8, p. 13.

Erlich, E.I., Sutherland, W.M., Hausel, W.D., and Zagruzina, I.A., 1989, Temporal distribution of the ultramafic-alkalic and alkalic rocks within the Russian, Siberian, and North American ancient platforms and their surroundings: Geological Survey of Wyoming Open File Report 89-9, 33 p.

Falk, R.W., 1992, Physical characteristics of diamonds from the George Creek, Kelsey Lake, and Chicken Park kimberlites: M.S. thesis, Colorado State University, Fort Collins, 188 p.

Finkelman, R.B., and Brown, R.D., 1989, Mineral resource and geochemical exploration potential of coal that has anomalous metal concentrations: U.S. Geological Survey Circular 1035, p. 18-19.

Fipke, C.E., Gurney, J.J., and Moore, R.O., 1995, Diamond exploration techniques emphasizing indicator mineral chemistry and Canadian examples: Geological Survey of Canada Bulletin 423, 86 p.

Forbes, R.B., Kline, J.T., and Clough, A.H., 1987, A preliminary evaluation of alluvial diamond discoveries in placer gravels of Crooked Creek, Circle district, Alaska: Alaska Division of Geological and Geophysical Surveys Report of Investigations 87-1, 26 p.

Foster, B.P., and Reitan, P.H., 1972, Kimberlite dike emplacement in the central Finger Lakes region, New York (abstract): Geological Society of America Abstracts with Programs, v. 4, p. 17.

Fuhrbach, J.R., 1992, Kilborne Hole peridot: Gems and Gemology, v. 28, p. 16-27.

Garrison, J.R., and Taylor, L.A., 1980, Megacrysts and xenoliths in kimberlite, Elliott County, Kentucky: a mantle sample beneath the Permian Appalachian plateau: Contributions to Mineralogy and Petrology, v. 75, p. 27-42.

Genth, F.A., 1891, The minerals of North Carolina: U.S. Geological Survey Bulletin 74, 119 p.

Gogineni, S.V., Melton, C.E., and Giardina, A.A., 1978, Some petrological aspects of the Prairie Creek diamond-bearing kimberlite diatreme, Arkansas: Contributions to Mineralogy and Petrology, v. 66, p. 251-266.

Goldich, S.S., Lidiak, E.G., Hedge, C.E., and Walthall, F.G., 1966, Geochronology of the midcontinent region–Part 2, northern area: Journal of Geophysical Research, v. 71, p. 5389-5408.

Graff, P.J., 1978, Geology of the lower part of the Early Proterozoic Snowy Range Supergroup, Sierra Madre, Wyoming, with chapters on Proterozoic regional tectonics and uraniferous quartz-pebble conglomerates: Ph.D. dissertation, University of Wyoming, Laramie, 85 p.

Griffin, W.L., and Ryan, C.G., 1993, Trace elements in garnets and chromites: evaluation of diamond exploration targets, *in* Diamonds: exploration, sampling, and evaluation, Proceedings of a short course: Prospectors and Developers Association of Canada, March 27, Toronto, Ontario, p. 185-212.

Gurney, J.J., 1986, A correlation between garnets and diamonds in kimberlites, *in* Glover, J.E., and Harris, P.G., editors, Kimberlite occurrence and origin: University of Western Australia Geology Department Publication 8, p. 143-166.

Gurney, J.J., 1989, Diamonds, *in* Ross, J., and others, editors, Kimberlites and related rocks, Volume 2, Their mantle/crust setting, diamonds and diamond exploration, Proceedings of the fourth international kimberlite conference, Perth, Australia, 1986: Geological Society of Australia Special Publication 14, p. 935-965.

Gurney, J.J., Levinson, A.A., and Smith, H.S., 1991, Marine mining of diamonds of the west coast of southern Africa: Gems and Gemology, v. 27, no. 4, p. 207-219.

Hall, G.M., and Amick, H.C., 1944, Igneous rock in the Norris region, Tennessee: Journal of Geology, v. 53, p. 424-430.

Hausel, W.D., 1981, Notes on diamond discussions with Dr. J.D. Love and Dr. R. Marrs, Wyoming State Geological Survey files, 1 p.

Hausel, W.D, 1991, Economic geology of the South Pass granite-greenstone belt, southern Wind River Range, western Wyoming: Wyoming State Geological Survey Report of Investigations 44, 125 p.

Hausel, W.D, 1993a, Preliminary report on the geology, geochemistry, mineralization, and mining history of the Seminoe Mountains mining district, Carbon County, Wyoming: Wyoming Geological Association Jubilee Anniversary Field Conference Guidebook, p. 387-409.

Hausel, W. D., 1993b, Metal and gemstone deposits of Wyoming, *in* Snoke, A.W., Steidtmann, J.R., and Roberts, S.M., editors, Geology of Wyoming:

Wyoming State Geological Survey Memoir 5, p. 816-835.

Hausel, W.D., 1994a, Mining history of Wyoming's gold, copper, iron, and diamond resources: Mining History Association 1994 Annual Meeting, Denver, Colorado, p. 27-44.

Hausel, W.D., 1994b, Pacific Coast diamonds–an unconventional source terrane: Wyoming State Geological Survey Mineral Report MR 94-8, 20 p.

Hausel, W.D., 1995a, Diamonds, kimberlites, lamproites, and related rocks in the United States: Exploration and Mining Geology, v. 4, no. 3, p. 243-270.

Hausel, W.D., 1995b, Diamonds and their host rocks in the United States: Mining Engineering, v. 47, no. 8, p. 723-732.

Hausel, W.D., 1996a, Recurring kimberlite and lamproite magmatism in the Wyoming craton–an overview (abstract): Geological Society of America Abstracts with Programs, 48th Annual Rocky Mountain Section, v. 28, no. 4, p. 10.

Hausel, W.D., 1996b, Overview of the Sloan 1 and 2 diamondiferous kimberlites, Colorado-Wyomimg State Line district, field trip guide, 1996 industrial minerals forum: Wyoming State Geological Survey Mineral Report MR 96-4, 14 p.

Hausel, W.D., 1996c, Pacific coast diamonds–an unconventional source terrane, *in* Coyner, A.R., and Fahey, P.L., editors, Geology and ore deposits of the American Cordillera: Geological Society of Nevada Symposium Proceedings, Reno, Nevada, p. 925-934.

Hausel, W.D., and Bond, J.E., 1994, Appalachian diamonds–old discoveries–new frontier: International California Mining Journal, v. 64, no. 3, p. 36-37.

Hausel, W.D., Edwards, B.R., and Graff, P.J., 1991, Geology and mineralization of the Wyoming Province: SME Preprint 91-72, 12 p.

Hausel, W.D., Glahn, P.R., and Woodzick, T.L., 1981, Geological and geophysical investigations of kimberlite in the Laramie Range of southeastern Wyoming: Wyoming State Geological Survey Preliminary Report 18, 13 p.

Hausel, W.D., and Gregory, R.W., 1996, Geophysical reconnaissance of the Lonetree circular topographic anomalies in the southern Green River Basin, Uinta County, Wyoming: Wyoming State Geological Survey Mineral Report MR 96-5, 11 p.

Hausel, W.D., Kucera, R.E., McCandless, T.E., and Gregory, R.W., 1996, Diamondiferous breccia pipes in the southern Green River Basin, Wyoming (abstract): 32nd Annual Forum on the Geology of Industrial Minerals, Laramie, Wyoming, Abstracts with Program, p. 23-24.

Hausel, W.D., Kucera, R.E., McCandless, T.E., and Gregory, R.W., 1997, Diamond exploration targets in the Wyoming craton, western United States: Wyoming State Geological Survey Mineral Report MR 97-2, 59 p.

Hausel, W.D., Love, C.M., and Sutherland, W.M., 1995, Road log–Leucite Hills, Green River Basin, Wyoming, *in* Jones, R.W., and Winter, G.A., editors, Wyoming Geological Association 1995 Field Conference Road Logs, Resources of Southwestern Wyoming: Wyoming Geological Association, Casper, Wyoming, p. 45-53.

Hausel, W.D., McCallum, M.E., and Roberts, J.T., 1985, The geology, diamond testing procedures, and economic potential of the Colorado-Wyoming kimberlite province–a review: Wyoming State Geological Survey Report of Investigations 31, 22 p.

Hausel, W.D., McCallum, M.E., and Woodzick, T.L., 1979, Exploration for diamond-bearing kimberlite in Colorado and Wyoming: an evaluation of exploration techniques: Wyoming State Geological Survey Report of Investigations 19, 29 p.

Hausel, W.D., and Stahl, S., 1995, The "Great Diamond Hoax of 1872": Wyoming Geological Association 1995 Field Conference Guidebook, Resources of Southwestern Wyoming, p. 13-27.

Hausel, W.D., Sutherland, W.M., and Gregory, E.B., 1988, Stream sediment sample results in search of kimberlite in southeastern Wyoming: Wyoming State Geological Survey Open File Report 88-11, 11 p.

Hausel, W.D., Sutherland, W.M., and Gregory, R.W., 1995, Lamproites, diamonds, indicator minerals, and related anomalies in the Green River Basin, Wyoming: Wyoming Geological Association 1995 Field Conference Guidebook, Resources of Southwestern Wyoming, p. 137-151.

Hearn, B.C., Jr., 1989, Alkalic ultramafic magmas in north-central Montana, U.S.A.: Genetic connections of alnöite, kimberlite, and carbonatite, *in* Ross, J., and others, editors, Kimberlites and related rocks, Volume 1, Their composition, occurrence, origin, and emplacement, Proceedings of the fourth international kimberlite conference, Perth, Australia, 1986: Geological Society of Australia Special Publication 14, p. 109-119.

Hearn, B.C., Jr., and McGee, E.S., 1983, Garnets in Montana diatremes: a key to prospecting for kimberlites: U.S. Geological Survey Bulletin 1604, 33 p.

Helmstaedt, H.H., 1993, Natural diamond occurrences and tectonic setting of primary diamond deposits, *in* Diamonds: exploration, sampling, and evaluation, Proceedings of a short course: Prospectors and Developers Association of Canada, March 27, Toronto, Ontario, p. 3-74.

Henage, L.F., 1972, A definitive study of the origin of lamproites: M.S. thesis, University of Oregon, Eugene, 83 p.

Heylum, E.B., 1985, Diamonds near Oroville, California: California Mining Journal, December, p. 63.

Hill, M., 1972, Hunting diamonds in California: Naturegraph Publishers, Healdsburg, California, 80 p.

Hills, F.A., and Houston, R.S., 1979, Early Proterozoic tectonics of the central Rocky Mountains, North America: University of Wyoming Contributions to Geology, v. 17, p. 89-110.

Hoffman, P.F., 1988, United plates of America, the birth of a craton: Early Proterozoic assembly and growth of Laurentia: Earth and Planetary Science, Annual Reviews, v. 16, p. 543.

Holden, R.J., 1944, The Punch Jones and other Appalachian diamonds: Bulletin of the Virginia Polytechnic Institute, v. 37, no. 4, 32 p.

Honess, A.P., and Graeber, C.K., 1924, A new occurrence of an igneous dike in southwestern Pennsylvania: American Journal of Science, v. 7, p. 313-315.

Houston, R.S., 1993, Late Archean and Early Proterozoic geology of southeastern Wyoming, *in* Snoke, A.W., Steidtmann, J.R., and Roberts, S.M., editors, Geology of Wyoming: Wyoming State Geological Survey Memoir 5, p. 78-116.

Janse, A.J.A., 1994a, Review of supposedly non-kimberlitic and non-lamproitic diamond host rocks, *in* Meyer, H.O.A., and Leonardos, O.H., editors, Diamonds: characterization, genesis, and exploration, Volume 2, Proceedings of the fifth international kimberlite conference, Araxa, Brazil, 1991: Companhia de Pesquisa de Recursos Minerais, Brazil, p. 144-159.

Janse, A.J.A., 1994b, Is Clifford's rule still valid? Affirmative examples from around the world, *in* Meyer, H.O.A., and Leonardos, O.H., editors, Diamonds: characterization, genesis, and exploration, Volume 2, Proceedings of the fifth international kimberlite conference, Araxa, Brazil, 1991: Companhia de Pesquisa de Recursos Minerais, Brazil, p. 215-235.

Jaques, A.L., Kerr, I.D., Lucas, H., Sun, S.S., and Chappell, B.W., 1989, Mineralogy and petrology of picritic monchiquites from Wandagee, Carnarvon Basin, Western Australia, *in* Ross, J., and others, editors, Kimberlites and related rocks, Volume 1, Their composition, occurrence, origin, and emplacement, Proceedings of the fourth international kimberlite conference, Perth, Australia, 1986: Geological Society of Australia Special Publication 14, p. 120-139.

Jaques, A.L., Lewis, J.D., and Smith, C.B, 1986, The kimberlites and lamproites of Western Australia: Geological Survey of Western Australia Bulletin 132, 268 p.

Johnson, R.W., Jr., and Milton, C., 1955, Dike rocks of central-western Virginia (abstract): Geological Society of America Bulletin, v. 66, p. 1689-1690.

Karlstrom, K.E., and Houston, R.S., 1984, The Cheyenne Belt: analysis of a Proterozoic suture in southern Wyoming: Precambrian Research, v. 25, p. 415-446.

Kay, S.M., Snedden, W.T., Foster, B.P., and Kay, R.W., 1983, Upper mantle and crustal fragments in the Ithaca kimberlites: Journal of Geology, v. 91, p. 277-290.

Kemp, J.F., and Ross, J.G., 1907, A peridotite dike in the coal measures of southwestern Pennsylvania: New York Academy of Sciences Annals, v. 17, p. 509-518.

Kidwell, A.L., 1947, Post-Devonian igneous activity in southeastern Missouri: Missouri Geological Survey Report of Investigations 4, 83 p.

Kidwell, A.L., 1990, Famous mineral localities, Murfreesboro, Arkansas: The Mineralogical Record, v. 21, p. 545-555.

Kirkley, M.B., Gurney, J.J., and Levinson, A.A., 1991, Age, origin, and emplacement of diamonds: Gems and Gemology, v. 27, no. 1, p. 2-25.

Kjarsgaard, B.A., 1996, Kimberlites, *in* LeCheminant, A.N., Richardson, D.G., DiLabio, R.N.W., and Richardson, K.A., editors, Searching for diamonds in Canada: Geological Survey of Canada Open File Report 3228, p. 29-37.

Koenig, J.B., 1956, The petrography of certain igneous rocks of Kentucky: Kentucky Geological Survey Bulletin 21, 57 p.

Kopf, R.W., Hurlbut, C.S., and Koivula, J.I., 1990, Recent discoveries of large diamonds in Trinity County, California: Gems and Gemology, v. 25, no. 3, p. 212-219.

Kreiger, W.E., Jr., and Thornton, C.P., 1976, Geology and petrology of the Two Buttes intrusion, southeast Colorado: Geological Society of America Abstracts with Programs, Rocky Mountain Section, p. 596-597.

Krol, L.G., 1988, Prairie Creek kimberlite (lamproite), *in* Colton, G.W., editor, Proceedings of the 22nd annual forum on the geology of industrial minerals: Arkansas Geological Commission Miscellaneous Publication MP-21, p. 73-75.

Kuehner, S.M., 1980, Petrogenesis of ultrapotassic rocks, Leucite Hills, Wyoming: M.S. thesis, University of Western Ontario, London, Ontario, 197 p.

Kunz, G.F., 1885, Precious stones, *in* Mineral resources of the United States: U.S. Geological Survey, p. 723-782.

Kunz, G.F., 1907, History of the gems found in North Carolina: North Carolina Geological and Economic Survey Bulletin 12, p. 5-9.

Lampietti, F.M.J., and Sutherland, D., 1978, Prospecting for diamonds–some current aspects: Mining Magazine, August, p. 117-123.

Larson, E.E., and Amini, M.H., 1981, Fission-track dating of the Green Mountain kimberlite diatreme near Boulder, Colorado: The Mountain Geologist, v. 18, no. 1, p. 18-22.

Laughlin, A.W., Charles, R.W., and Aldrich, M.J., Jr., 1989, Heteromorphism and crystallization paths of katungites, Navajo volcanic field, Arizona, U.S.A., *in* Ross, J., and others, editors, Kimberlites and related rocks, Volume 1, Their composition, occurrence, origin, and emplacement, Proceedings of the fourth international kimberlite conference, Perth, Australia, 1986: Geological Society of Australia Special Publication 14, p. 582-591.

Levinson, A.A., Gurney, J.J., and Kirkley, M.B., 1992, Diamond sources and production: past, present, and future: Gems and Gemology. v. 28, no. 4, p. 234-254.

Lopez, D.A., 1995, Geology of the Sweet Grass Hills, north-central Montana: Montana Bureau of Mines and Geology Memoir 68, 31 p.

Love, J.D., 1963, Letter to W.H. Bradley dated April 25, 1963: Wyoming Geological Survey files, 6 p.

Lush, A.P., McGrew, A.J., Snoke, A.W., and Wright, J.E., 1988, Allochthonous Archean basement in the northeastern East Humboldt Range, Nevada: Geology, v. 16, p. 349-353.

Mansker, W.L., 1973, Petrology of southeastern Missouri ultramafic pipe: M.S. thesis, University of Missouri, Rolla, 53 p.

Mansker, W.L., Richards, B.D., and Cole, G.P., 1987, A note on newly discovered kimberlites in Riley County, Kansas, *in* Morris, E.M., editor, Mantle metasomatism and alkaline magmatism: Geological Society of America Special Paper 215, p. 197-204.

Markezich, M.A., 1985, A geophysical study of the Hill's Pond lamproite, Woodson and Wilson Counties, Kansas: M.S. thesis, University of Kansas, Lawrence, 57 p.

Marks, J.E., 1985, Multispectral remote sensing techniques applied to exploration for kimberlite diatremes, Laramie Range, Wyoming-Colorado: M.S. thesis, University of Wyoming, Laramie, 164 p.

Martens, J.H.C., 1924, Igneous rocks of Ithaca, New York and vicinity: Geological Society of America Bulletin, v. 35, p. 305-320.

Marrs, R.W., Marks, J.E., Hausel, W.D., and Albert, K.E., 1984, Detection of diamond-bearing kimberlites in the Colorado/Wyoming Province: University of Wyoming Remote Sensing Laboratory Open File Report, 70 p.

Mather, B., 1941, Diamonds in North America: Natural History Society of Maryland Bulletin, v. 11, no. 6, p. 106-112.

McCallum, M.E., 1991, The Sloan 1 and 2 kimberlite complex near the southern boundary of the State

Line district of the Colorado-Wyoming kimberlite province: Wyoming Geological Association 42nd Field Conference Guidebook, Mineral Resources of Wyoming, p. 229-250.

McCallum, M.E., and Eggler, D.H., 1979, Field guide for the Sloan and Nix kimberlites in the southern portion of the Colorado-Wyoming State Line kimberlilte district, *in* Ethridge, F.G., editor, Field guide to the northern Front Range and northeastern Denver Basin, Colorado: Geological Society of America 32nd Annual Field Conference Guidebook, p. 181-209.

McCallum, M.E., Eggler, D.H., and Burns, L.K., 1975, Kimberlitic diatremes in northern Colorado and southern Wyoming, *in* Physics and chemistry of the Earth, v. 9: Pergamon Press, New York, p. 149-161.

McCallum, M.E., Eggler, D.H., Coopersmith, H.G., Smith, C.B., and Mabarak, C.D., 1977, Field guide, Colorado-Wyoming State Line district: Second international kimberlite conference, Santa Fe, New Mexico, 23 p.

McCallum, M.E., and Mabarak, C.D., 1976, Diamond in State Line kimberlite diatremes, Albany County, Wyoming, and Larimer County, Colorado: Wyoming State Geological Survey Report of Investigations 12, 36 p.

McCallum, M.E., Mabarak, C.D., and Coopersmith, H.G., 1979, Diamonds from kimberlites in the Colorado-Wyoming State Line district, *in* Boyd, F.R., and Meyer, H.O.A., editors, Kimberlites, diatremes, and diamonds: Their geology, petrology, and geochemistry, Volume 1, Proceedings of the second international kimberlite conference: American Geophysical Union, Washington, D.C., p. 42-58.

McCallum, M.E., and Waldman, M.A., 1991, The diamond resources of the Colorado-Wyoming State Line district: kimberlite indicator mineral chemistry as a guide to economic potential: Wyoming Geological Association 42nd Field Conference Guidebook, Mineral Resources of Wyoming, p. 77-90.

McCandless, T.E., 1982, The mineralogy, morphology, and chemistry of detrital minerals of a kimberlitic and eclogitic nature, Green River Basin, Wyoming: M.S. thesis, University of Utah, Salt Lake City, 107 p.

McCandless, T.E., 1984, Detrital minerals of mantle origin in the Green River Basin, Wyoming: Society of Mining Engineers of AIME Reprint 84-395, 6 p.

McCandless, T.E., and Gurney, J.J., 1989, Sodium in garnet and potassium in clinopyroxene: criteria for classifying mantle eclogites *in* Ross, J., and others, editors, Kimberlites and related rocks, Volume 2, Their mantle/crust setting, diamonds and diamond exploration, Proceedings of the fourth international kimberlite conference, Perth, Australia, 1986: Geological Society of Australia Special Publication 14, p. 827-832.

McCandless, T.E., Nash, W.P., and Hausel, W.D., 1995, Mantle indicator minerals in ant mounds and conglomerates of the southern Green River Basin, Wyoming: Wyoming Geological Association 1995 Field Conference Guidebook, Resources of Southwestern Wyoming, p. 153-164.

McGretchin, T.R., 1968, The Moses Rock dike: geology, petrology, and mode of emplacement of kimberlite-bearing breccia dike, San Juan County, Utah: Ph.D. thesis, California Institute of Technology, Pasadena, 188 p.

McGretchin, T.R., and Silver, L.T., 1970, Compositional relations in minerals from kimberlite and related rocks in the Moses Rock dike, San Juan County, Utah: American Mineralogist, v. 55, p. 1738-1771.

Meen, J.K., Eggler, D.H., and McCallum, M.E., 1986, Proterozoic granulite xenoliths from Ming Bar diatreme, Montana: Geological Society of America Abstracts with Programs, v. 18, p. 692.

Meyer, H.O.A., 1975, Kimberlite from Norris Lake, eastern Tennessee: Mineralogy and petrology: Journal of Geology, v. 83, no. 4, p. 518-526.

Meyer, H.O.A., 1976, Kimberlites of the Continental United States: The Journal of Geology, v. 84, no. 4, p. 377-403.

Meyer, H.O.A., Lewis, R.D., Bolivar, S., and Brookins, D.C., 1977, Prairie Creek kimberlite, Murfreesboro, Pike County, Arkansas–general description for field excursion: Second international kimberlite conference, Santa Fe, New Mexico, 17 p.

Meyer, H.O.A., and Kridelbaugh, S.J., 1977, Green Mountain kimberlite, Colorado (abstract): Second international kimberlite conference extended abstracts, Santa Fe, New Mexico, p. 224.

Miser, H.D., 1914, New areas of diamond-bearing peridotite in Arkansas: U.S. Geological Survey Bulletin 540, p. 534-546.

Miser, H.D., and Ross, C.S., 1922a, Diamond-bearing peridotite in Pike County, Arkansas: U.S. Geological Survey Bulletin 735-I, p. 279-322.

Miser, H.D., and Ross, C.S., 1922b, Peridotite dikes in Scott County, Arkansas: U.S. Geological Survey Bulletin 735-I, p. 271-278.

Mitchell, R.H., 1986, Kimberlites: mineralogy, geochemistry, and petrology: Plenum Press, New York, New York, 442 p.

Mitchell, R.H., 1995, Kimberlites, orangeites, and related rocks: Plenum Press, New York, New York, 410 p.

Mitchell, R.H., and Bergman, S.C., 1991, Petrology of lamproites: Plenum Press, New York, New York, 447 p.

Naeser, C.W., 1971, Geochronology of the Navajo-Hopi diatremes, Four Corners area: Journal of Geophysical Research, v. 76, no. 20, p. 4978-4985.

Naeser, C.W., and McCallum, M.E., 1977, Fission-track dating of kimberlitic zircons (abstract): Second international kimberlite conference, extended abstracts, Santa Fe, New Mexico, p. 242-243.

Ogden, P.R., Jr., 1979, The geology, major element geochemistry, and petrogenesis of the Leucite Hills volcanic rocks, Wyoming: Ph.D. dissertation, University of Wyoming, Laramie, 137 p.

Oliver, D.H., 1990, Prairie Divide properties corporation, Sloan Ranch diamond project, Larimer County, Colorado, summary report: Unpublished Consultant's Report, Wyoming State Geological Survey files, 11 p.

Otter, M.L., and Gurney, J.J., 1989, Mineral inclusions in diamonds from the Sloan diatremes, Colorado-Wyoming State Line kimberlite district, North America, *in* Ross, J., and others, editors, Kimberlites and related rocks, Volume 2, Their mantle/crust setting, diamonds and diamond exploration, Proceedings of the fourth international kimberlite conference, Perth, Australia, 1986: Geological Society of Australia Special Publication 14, p. 1042-1053.

Parrish, J.B., 1984, Tectonic model for kimberlite exploration in the Appalachians: Society of Mining Engineers of AIME Preprint 84-397, 4 p.

Parrish, J.B., and Lavin, P.M., 1982, Tectonic model for kimberlite emplacement in the Appalachian Plateau of Pennsylvania: Geology, v. 10, p. 344-347.

Paterson, N.R., and MacFadyen, D.A., 1984, An airborne EM (INPUT)/magnetometer survey, State Line district, Colorado-Wyoming: Society of Mining Engineers of AIME Preprint 84-310, 11 p.

Paydirt, 1996, Colorado diamond weighs in at 28.3 carats: Pay Dirt, September, 1996, p. 10.

Pearson, D.G., Davies, G.R., and Nixon, P.H., 1993, Geochemical constraints on the petrogenesis of diamond facies pyroxenites from the Beni Bousera peridotite massif, North Morocco: Journal of Petrology, v. 34, pt. 1, p. 125-172.

Pearson, D.G., Davies, G.R., Nixon, P.H., and Milledge, H.J., 1989, Graphitized diamonds from a peridotite massif in Morocco and implications for anomalous diamond occurrences: Nature, v. 338, p. 60-62.

Peterman, Z.E., Hedge, C.E., and Braddock, W.A., 1968, Age of Precambrian events in the northeast Front Range, Colorado: Journal of Geophysical Research, v. 73, p. 2277-2296.

Peterson, T.D., 1996, Lamproites, *in* LeCheminant, A.N., Richardson, D.G., DiLabio, R.N.W., and Richardson, K.A., editors, Searching for diamonds in Canada: Geological Survey of Canada Open File Report 3228, p. 79-86.

Raeside, R.P., and Helmstaedt, H., 1982, The Ile Bizard intrusion, Montreal, Quebec–kimberlite or lamprophyre?: Candian Journal of Earth Sciences, v. 19, no. 10, p. 1996-2011.

Rock, N.M.S., Griffin, B.J., Edgar, A.D., Paul, D.K., and Hergt, J.M., 1992, A spectrum of potentially diamondiferous lamproites and minettes from the Jharia coalfield, eastern India: Journal of Volcanology and Geothermal Research, v. 50, p. 55-83.

Roden, M.F., 1981, Origin of coexisting minette and ultramafic breccia, Navajo volcanic field: Contributions to Mineralogy and Petrology, v. 77, p. 195-206.

Roden, M.F., and Smith, D., 1979, Field geology, chemistry, and petrology of Buell Park minette diatreme, Apache County, Arizona, *in* Boyd, F.R., and Meyer, H.O.A., editors, Kimberlites, diatremes, and diamonds: Their geology, petrology, and geochemistry, Volume 1, Proceedings of the second international kimberlite conference: American Geophysical Union, Washington, D.C., p. 364-381.

Rogers, J.A., Jr, 1985, Kimberlite exploration, Red Feather area, and petrology of the Chicken Park diatreme, northern Colorado: M.S. thesis, Colorado State University, Fort Collins, 196 p.

Rosa, F., and Brookins, D.G., 1967, The mineralogy of the Stockdale kimberlite pipe, Riley County, Kansas: Transactions of the Kansas Academy of Science, v. 69, nos. 3-4, p. 335-344.

Rosenhouse, L., 1975, Diamonds along the Feather River: Gems and Minerals, October, 1975, p. 36-55.

Salpas, P.A., Taylor, L.A., and Shervais, J.W., 1986, The Blue Ball, Arkansas kimberlite: mineralogy, petrology, and geochemistry: Journal of Geology, v. 94, p. 891-901.

Scambos, T.A., 1991, Isotopic and trace-element characteristics of the central Montana alkalic province kimberlite-alnöite suite, *in* Baker, D.W., and Berg, R.B., editors, Guidebook of the central Montana alkalic province–Geology, ore deposits, and origins: Montana Bureau of Mines and Geology Special Publication 100, p. 93-109.

Schmitt, H.H., Swann, G.A., and Smith, D., 1974, The Buell Park kimberlite, northeastern Arizona, *in* Karlstrom, T.N., Swann, G.A., and Eastwood, R.L., editors, Geology of northern Arizona: Geological Society of America, Flagstaff, Arizona, p. 672-698.

Schulze, D.J., 1985, Evidence for primary kimberlitic liquids in megacrysts from kimberlites in Kentucky, U.S.A.: Journal of Geology, v. 93, p. 75-79.

Schulze, D.J., 1992, Diamond eclogite from Sloan Ranch, Colorado, and its bearing on the diamond grade of the Sloan kimberlite: Economic Geology, v. 87, no. 8, p. 2175-2179.

Schulze, D.J., 1993, Garnet xenocryst populations in North American kimberlites, *in* Diamonds: exploration, sampling, and evaluation, Proceedings of a short course: Prospectors and Developers Association of Canada, March 27, Toronto, Ontario, p. 359-378.

Sears, C.E., and Gilbert, M.C., 1973, Petrography of the Mt. Horeb Virginia kimberlite (abstract): Geological Society of America Abstracts with Programs, v. 5, p. 434.

Shannon, E.V., 1926, Minerals of Idaho: U.S. National Museum Bulletin 131, p. 16.

Shaver, K.C., 1988, Exploration of the Sloan Ranch complex: a diamondiferous kimberlite prospect in northern Colorado: Mining Engineering, v. 40, no.1, p. 45-48.

Shaver, K.C., 1994, Phase 1 underground exploration and mill testing of the Sloan Ranch diamondiferous kimberlite complex, Larimer County, Colorado: Royalstar Resources Ltd, company report (unpublished), 39 p.

Sims, P.K., 1995, Archean and Early Proterozoic tectonic framework of north-central United States and adjacent Canada: U.S. Geological Survey Bulletin 1904, 12 p.

Sinkankas, J., 1959, Gemstones of North America: Van Nostrand Company, New York, New York, 675 p.

Sinkankas, J., 1964, Mineralogy: Van Nostrand Company, New York, New York, 585 p.

Smith, C.B., 1977, Kimberlite and mantle derived xenoliths at Iron Mountain, Wyoming: M.S. thesis, Colorado State University, Fort Collins, 218 p.

Smith, C.B., 1979, Rb-Sr mica ages on various kimberlites (abstract): Kimberlite Symposium II, Cambridge, England, p. 61-64.

Smith, C.B., 1983, Rb-Sr, U-Pb, Sm-Nd isotopic studies of kimberlite and selected mantle-derived xenoliths: Ph.D. dissertation, University of Witwatersrand, Johannesburg, South Africa, 436 p.

Smith, D., 1979, Hydrous minerals and carbonates in peridotitic inclusions from the Greens Knob and Buell Park kimberlite diatremes on the Colorado Plateau, *in* Boyd, F.R., and Meyer, H.O.A., editors, Kimberlites, diatremes, and diamonds: Inclusions in kimberlites and other volcanics, Volume 2, Proceedings of the second international kimberlite conference: American Geophysical Union, Washington, D.C., p. 213-224.

Smith, D., in press, Chlorite-rich ultramafic reaction zones in Colorado Plateau xenoliths: recorders of sub-Moho hydration: Contributions to Mineralogy and Petrology, 28 p.

Smithson, S.B., 1959, The geology of the southeastern Leucite Hills, Sweetwater County, Wyoming: M.A. thesis, University of Wyoming, Laramie, 92 p.

Sosman, R.H., 1938, Evidence on the intrusion-temperature of peridotites: American Journal of Science, v. 35, p. 353-359.

Southworth, C.S., Gray, K.J., and Sutter, J.F., 1992, Middle Eocene intrusive igneous rocks of the central Appalachian Valley and Ridge Province–setting, chemistry, and implications of crustal structure: U.S. Geological Survey Bulletin 1839, 24 p.

Speer, J.T., 1985, Xenoliths of the Leucite Hills volcanic rocks, Sweetwater County, Wyoming: M.S. thesis, University of Wyoming, Laramie, 57 p.

Sweet, P.C., 1997, Diamonds in Virginia, *in* Jones, R.W., and Harris, R.E., editors, Proceedings of the 32nd annual forum on the geology of industrial minerals: Wyoming State Geological Survey Public Information Circular 38, p. 269-276.

Taylor, L.A., 1984, Kimberlitic magmatism in the eastern United States: relationships to Mid-Atlantic tectonism, *in* Kornprobst, J., editor, Kimberlites and related rocks: Elsevier Publications, Amsterdam, The Netherlands, p. 417-424.

U.S. Geological Survey, 1968, Mineral resources of the Appalachian region: U.S. Geological Survey Professional Paper 580, 301 p.

Wagner, P.A., 1914, The diamond fields of southern Africa: The Transvall Leader, Johannesburg, South Africa, 347 p.

Whitaker, M.C., 1898, An olivinite dike of the Magnolia district and the associated picrotitanite: Colorado Scientific Society Proceedings, p. 104-118.

Williams, H., Rivers, T., Hoffman, P.F., Monger, J.W., and Lewry, J.F., 1986, The making of North America (abstract): Geological Society of America Abstracts with Programs, v. 18, p. 789.

Wojcik, K.M., and Knapp, R.W., 1990, Stratigraphic control of the Hill's Pond lamproite, Silver City Dome, southeastern Kansas: Geology, v. 18, p. 251-254.

Zartman, R.E., Brock, M.R., Heyl, A.V., and Thomas, H.H., 1967, K-Ar and Rb-Sr ages of some alkalic intrusive rocks from central and eastern United States: American Journal of Science, v. 265, p. 848-870.